BRUSH UP YOUR MATH

A HOW-TO HANDBOOK

BRUSH UP YOUR MATH

A HOW-TO HANDBOOK

BETTY WOODSIDE

McGraw-Hill Ryerson Limited
Toronto Montréal New York St. Louis San Francisco
Aukland Bogotá Guatemala Hamburg
Lisbon London Madrid Mexico New Delhi Panama
Paris San Juan São Paulo Singapore Sydney Tokyo

BRUSH UP YOUR MATH: A HOW-TO HANDBOOK

ISBN 0-07-092375-2

3 4 5 6 7 8 9 10 D 10 9 8 7

Printed and bound in Canada

Woodside, Betty, date.
Brush Up Your Math

ISBN 0-07-092375-2

1. Mathematics – Programmed instruction.
2. Mathematics -- 1961-   I. Title.

QA39.2.W66   510'.07'7   C81-094559-2

Table of Contents

Preface

Introduction – How to Use this Book

PREFACE

Brush Up Your Math is designed for those who would like to improve their math skills and for those who may be preparing for various college pre-admission tests.

The book is made to be both enjoyable and practical — perfect for retired people, students, or those who, for the sheer enjoyment of it, would like to "brush up their math."

Since many pre-admission tests now call for competence in basic math, **Brush Up Your Math** ensures proper preparation for adults who are returning "to school" for retraining, pleasure, or to acquire new skills.

For students who are "brushing up their math" this book suggests:
(a) what they can expect on a math test
(b) what their own weak areas may be, and
(c) material to deal effectively with any math problems.

A variety of tests are presented dealing with certain fundamental math operations; a preliminary self-evaluation test is given to determine any weaknesses; and students are referred to pages of exercises in order to help them overcome any problems they might have.

The building-block sequence ensures that students and others employing this self-study guide can enjoy success by closely following instructions for the book's use. Each chapter offers:
1. Introduction
2. Objectives
3. Method of solution spelled out in numbered steps
4. Examples
5. Practice exercises with answers
6. End of chapter tests

Brush Up Your Math has been divided into two parts. Mastery of Part One is sufficient for most programs that require Grade 10 or equivalent. Part Two is designed for programs with a heavier math emphasis. An appendix contains supplementary exercises in both parts.

The author would like to acknowledge the support, advice, and assistance of the following people in the planning and development of this book:

Jim Carswell   Ken MacLennan
Terry Joy   Mac Mackay
the late Jane Martin   Bill Rapson

HOW TO USE THIS BOOK

First: Find out whether your particular goal requires mastery of Part One only, or Parts One and Two both.

Second: Do the preliminary test for Part One, and mark your answers right or wrong.

Third: List the code numbers for all the wrong answers. Examine each incorrect answer. If the error is obvious, correct it, and cross that number off your list.

Fourth: If you are unable to correct an error, study the section of the book that has the same code number. Be sure to work out the examples and do the practice exercises. When you feel sure that you have mastered the particular problem that caused your wrong answer, go back to the test item and re-work it. If the answer is correct, cross that item off your list. Continue to work systematically through the list of errors until there are none left.

Fifth: Do the post test for Part One in order to confirm that you really have solved your problems.

Sixth: If Part Two is required, follow the same procedure.

If you had just a few wrong answers in the preliminary test, you should be able to review the work quickly, in a matter of hours or days. If you had quite a few wrong answers, it may take you several weeks to review. If most of your answers were incorrect, you would be well-advised to seek help rather than trying to review on your own.

Very few people can "learn" math by merely reading or looking at examples. If you copy the question, then work out each step and compare your work with the example, you will be able to focus on the exact spot where the difficulty occurs.

Much of the time, you will find answers on the same page as the question. Use the answers freely, on a one-by-one basis if you like. You will then have immediate reassurance that you are on the right track, or a warning that some point still has to be cleared up. The exceptions to this are answers to tests. We suggest that you complete an entire test before checking answers.

END-OF-CHAPTER TESTS

The end-of-chapter tests that are used in Part One of this book were given to several thousand adults as part of a pre-admission test to skill programs at a community college.

Two surveys, conducted three years apart, yielded the data given below. You may wish to compare your scores with the averages. If you do so, you should realize that in many cases the scores obtained were not satisfactory for direct entry to a skill program.

AVERAGE TEST RESULTS

Section of test	Average score
Whole numbers:	77.8%
Addition	91.4%
Subtraction	86.2%
Multiplication	69.8%
Division	64.0%
Fractions	48.0%
Decimals	43.9%
Percent	31.6%

To find your percent, form a fraction and multiply by 100%.

$$\frac{\text{The number of your correct answers}}{\text{The number of possible correct answers}} \times 100\%$$

Example: You had 12 correct answers in fractions.

$$\frac{12}{16} \times 100\% = 75\%. \text{ Your score would be } 75\%.$$

Answers for these tests are on pages 118 and 119.

PRELIMINARY TEST — PART ONE

Complete the entire test (or as much as you are able to do) before referring to the answers. Then proceed according to the instructions on page vii.

WHOLE NUMBERS

COMMON FRACTIONS

DECIMALS

PAGE		
60	**D – 1**	16.0021 is equal to which of the following?

D – 1 16.0021 is equal to which of the following?

    (a) sixteen and 21 hundredths   (b) sixteen and 21 ten-thousandths

    (c) sixteen and 21 millionths   (d) sixteen point two one thousandths

61

D – 2 Write numerals for the following:

    (a) four thousand sixty-five ten-thousandths

    (b) four thousand and sixty-five ten-thousandths

63

D – 3 Write 0.0025 as a common fraction in lowest terms.

64

D – 4 Add or subtract as indicated.

    (a) $0.2 + 0.5 + 0.3$ (b) $0.24 + 6.7 + 1 + 0.009$

    (c) $13 - 0.07$ (d) 5 hundredths $-$ 5 thousandths

66

D – 5 Multiply.

    (a) 1.25×400 (b) $0.2 \times 0.2 \times 0.2$ (c) 2.3 by 0.04

68

D – 6 Divide.

    (a) 0.05 by 25   (b) $0.03\overline{)600}$ (c) $\frac{4}{15}$ correct to 2 places

72

D – 7 Complete the following:

    (a) To multiply by 1000, move the decimal point ___ places to the ____

    (b) To divide by 100, move the decimal point ___ places to the ____.

    (c) $0.42 \div 1000 =$ ___   (d) $16.43 \times 10\,000 =$ ___.

74

D – 8 Round (off) as indicated.

    (a) 0.143 to the nearest tenth

    (b) 0.06527 to 3 places

    (c) 421.692 to the nearest whole number

77

D – 9 Change the following to decimal form: $\frac{3}{16}, 2\frac{11}{12}, 16\frac{1}{40}$.

PERCENT

	PAGE

P – 1 Change to fractions in lowest terms:　　　　　　　　　86
　　　(a) 40%　　(b) $\frac{2}{3}$%　　(c) $6\frac{1}{4}$%　　(d) $212\frac{1}{2}$%

P – 2 Change to decimal form: (a) 125%　　(b) 0.2%　　(c) $10\frac{3}{4}$%　　88

P – 3 Change to percent equivalents:　　　　　　　　　　89
　　　(a) $\frac{3}{5}$　　(b) 7.5　　(c) $1\frac{1}{12}$　　(d) 0.1325

P – 4 Calculate the following:　　　　　　　　　　　92
　　　(a) 2% of 1400　　　　(b) 600% of 12
　　　(c) 3.75% of 180　　　(d) $\frac{1}{4}$% of 4000

P – 5 Express the first number in each pair as a percent of the second.　95
　　　(a) 13 and 20　　(b) $\frac{1}{4}$ and $2\frac{1}{2}$　　(c) $13.25 and $212.00

P – 6 In each of the following, find the value of 100%:　　　97
　　　(a) if 3% of a number is 12
　　　(b) if 8.5% of a number is 12.41
　　　(c) if 0.09% of a number is 27

P – 7 A proportion is a statement that two _____ are _____.　106

P – 8 What is the value of a in the proportion, $a:35 = 21:49$?　107

P – 9 Rearrange the following proportion to show ratios rather than rates:　110
　　　$9:45 = 63:315$ represents 9 km in 45 min compared with 63 km in 315 min.

ANSWERS FOR PRELIMINARY TEST — PART ONE

WHOLE NUMBERS

W – 1 2 is millions, 6 is tens of thousands, 0 is hundreds.

W – 2 Three billion, sixteen million, four hundred eight

W – 3 10 000 600 045

W – 4 Multiplication and addition are both operations that combine.

W – 5 (a) The average is $\dfrac{22 + 37 + 17 + 16}{4} = \dfrac{92}{4} = 23$

　　　(b) 7^2　　(c) $6^4 = 6 \times 6 \times 6 \times 6$ or 1296

　　　(d) The square root of 81 = 9　　(e) 20

W – 6 (a)　　　23 765　　(b)　32 004　　(c)　　　6049　　(d)　　　20 003
　　　　　　　　900　　　　　−27 968　　　　　×4080　　　24) 480 072
　　　　　　87 659　　　　　　4 036　　　　　483920　　　　　48
　　　　　　　9 946　　　　　　　　　　　　241960　　　　　00 072
　　　　　　+123 497　　　　　　　　　　　24679920　　　　　　72
　　　　　　245 767　　　　　　　　　　　　　　　　　　　　　　0

W – 7 (a)   *Subtraction*   *Proof*   (b) *Multiplication*   *Proof*

$$\begin{array}{r} 4000 \\ -2008 \\ \hline 1992 \end{array} \qquad \begin{array}{r} 2008 \\ +1992 \\ \hline 4000 \end{array} \qquad\qquad \begin{array}{r} 270 \\ \times 300 \\ \hline 81\,000 \end{array}$$

$$300\overline{)81000} \\ \underline{600} \\ 2100 \\ \underline{2100} \\ 0$$

(c) *Division*   *Proof*

$$360\overline{)86400} \\ \begin{array}{r} 240 \\ \end{array}$$

$$\begin{array}{r} 720 \\ \hline 1440 \\ \underline{1440} \\ 0 \end{array} \qquad \begin{array}{r} 240 \\ \times 360 \\ \hline 14400 \\ 720 \\ \hline 86400 \end{array}$$

COMMON FRACTIONS

F – 1 (a) any fraction that has 7 as the upper part, eg. $\dfrac{7}{16}$

(b) $12\dfrac{2}{3} = \dfrac{3 \times 12 + 2}{3} = \dfrac{38}{3}$

F – 2 (a) $\dfrac{75}{120} = \dfrac{75 \div 15}{120 \div 15} = \dfrac{5}{8}$ (b) $\dfrac{21 \div 7}{28 \div 7} = \dfrac{3}{4}$ and $\dfrac{48 \div 16}{64 \div 16} = \dfrac{3}{4}$

F – 3 $11\dfrac{2}{3} + 4\dfrac{5}{16} + 9\dfrac{5}{8} + 8 + 7\dfrac{7}{12} = 39\dfrac{32 + 15 + 30 + 28}{48}$

$$= 39\dfrac{105}{48} = 41\dfrac{9}{48} = 41\dfrac{3}{16}$$

F – 4 $12\dfrac{2}{15} \\ -\ 5\dfrac{3}{4} \end{} = \begin{} 12\dfrac{8}{60} \\ -\ 5\dfrac{45}{60} \end{} = \begin{} 11\dfrac{68}{60} \\ -\ 5\dfrac{45}{60} \\ \hline 6\dfrac{23}{60} \end{}$

F – 5 $1\dfrac{1}{5} \times \dfrac{6}{11} \times 2\dfrac{1}{4} \times 15 \times 2\dfrac{4}{9} = \dfrac{6}{5} \times \dfrac{6}{11} \times \dfrac{9}{4} \times \dfrac{15}{1} \diagup \dfrac{22}{9} = \dfrac{54}{1} = 54$

F – 6 (a) $14 \div \dfrac{1}{2} = \dfrac{14}{1} \times \dfrac{2}{1} = \dfrac{28}{1} = 28$ (b) $\dfrac{3}{4} \div \dfrac{12}{1} = \dfrac{3}{4} \times \dfrac{1}{12} = \dfrac{1}{16}$

(c) $1\dfrac{1}{6}$

F – 7 42 is $\dfrac{42}{48}$ of 48, or $\dfrac{7}{8}$ of 48

F – 8 The number is 56.

DECIMALS

D – 1 16.0021 = sixteen and 21 ten-thousandths

D – 2 (a) 0.4065   (b) 4000.0065 (the decimal point is read "and")

D – 3 $0.0025 = \dfrac{25}{10\ 000} = \dfrac{1}{400}$

D – 4 (a) 0.2   (b) 0.24   (c) 13.00   (d) 0.050
            0.5   6.7   0.07   0.005
            0.3   1   _____   0.045
           ____   0.009   12.93
            1.0   _____
                             7.949

D – 5 (a) 500   (b) 0.008   (c) 0.092

D – 6 (a) 0.002   (b) 20 000   (c) 0.27

D – 7 (a) To multiply by 1000, move the decimal point 3 places to the right.
        (b) To divide by 100, move the decimal point 2 places to the left.
        (c) $0.42 \div 1000 = 0.000\ 42$ (d) $16.43 \times 10\ 000 = 164\ 300$

D – 8 (a) 0.143 to the nearest tenth is 0.1
        (b) 0.06527 to 3 places is 0.065
        (c) 421.692 to the nearest whole number is 422

D – 9 0.1875; 2.91$\dot{6}$ or 2.917 (rounded to 3 places); 16.025

PERCENT

P – 1 (a) $40\% = \dfrac{2}{5}$ (b) $\dfrac{2}{3}\% = \dfrac{1}{150}$ (c) $6\dfrac{1}{4}\% = \dfrac{1}{16}$ (d) $212\dfrac{1}{2}\% = 2\dfrac{1}{8}$

P – 2 (a) $125\% = 1.25$ (b) $0.2\% = 0.002$ (c) $10\dfrac{3}{4}\% = 0.1075$

P – 3 (a) $\dfrac{3}{5} = 60\%$ (b) $7.5 = 750\%$

        (c) $1\dfrac{1}{12} = 108\dfrac{1}{3}\%$ (d) $0.1325 = 13.25\%$

P – 4 (a) 2% of 1400 = 28   (b) 600% of 12 = 72   (c) 6.75   (d) 10

P – 5 (a) 13 is 65% of 20   (b) $\dfrac{1}{4}$ is 10% of $2\dfrac{1}{2}$ (c) $13.25 is 6.25% of $212.00

P – 6 (a) 3% of 400 is 12.   (b) 8.5% of 146 is 12.41   (c) 0.09% of 30 000 is 27.

P – 7 ... that two ratios are equal

P – 8 $a = 15$

P – 9 Proportion composed of ratios: 9:63 = 45:315

CHAPTER ONE: **WHOLE NUMBERS**

Whole numbers include the natural or counting numbers such as 1, 2, 3, 4 etc., as opposed to the rational numbers such as fractions, decimals and percent.

This chapter deals with the vocabulary relating to whole numbers (and to arithmetic in general) and provides practice in the fundamental operations of addition, subtraction, multiplication and division.

It is assumed that a student who is able to profit from a self-help book such as this one, will have at least a rudimentary skill in these four operations. The emphasis, therefore, is on self assessment and error awareness, rather than how-to-do-it explanations.

ITEMS IN CHAPTER

ITEM W – 1 PLACE VALUE IN WHOLE NUMBERS

Our number system is called the decimal number system because its base is ten. (The Latin word *decem* means ten). Each digit in a numeral refers to a specific power of 10. Each digit has a "place value"; that is, its value is determined by its place or location in relation to the decimal point.

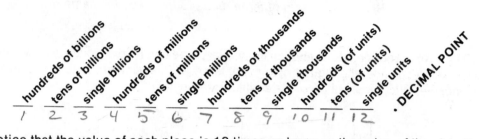

Notice that the value of each place is 10 times as large as the value of the place to its immediate right, and 10 times as small as the value of the place to its immediate left.

The numeral 28 465 092 has a 2 in the first place and a 2 in the eighth place. The value of the one is ten million times as great as the value of the other.

The numeral 62 038 751 is read:

sixty-two million, thirty-eight thousand, seven hundred fifty-one.

It is composed of:

1 unit	1	$(1 \times 1$ or $1 \times 10^0)$
5 tens	50	$(5 \times 10$ or $5 \times 10^1)$
7 hundreds	700	$(7 \times 100$ or $7 \times 10^2)$
8 single thousands	8 000	$(8 \times 1000$ or $8 \times 10^3)$
3 tens of thousands	30 000	$(3 \times 10\,000$ or $3 \times 10^4)$
0 hundreds of thousands	000 000	$(0 \times 100\,000$ or $0 \times 10^5)$
2 single millions	2 000 000	$(2 \times 1\,000\,000$ or $2 \times 10^6)$
6 tens of millions	60 000 000	$(6 \times 10\,000\,000$ or $6 \times 10^7)$
	62 038 751	

EXERCISE W – 1

1. What is the place value of each underlined digit in the following:
   (a) 324<u>6</u>   (b) 7 3<u>5</u>6 284   (c) 6<u>0</u>0 02<u>7</u>   (d) <u>6</u> 4<u>0</u>0 000
2. Write the numeral that has
   2 hundreds of thousands, 6 units, 9 hundreds, 7 tens of thousands, 0 single thousands and 5 tens.   *270956*

ANSWERS FOR EXERCISE W – 1
1. (a) 3 (thousands) 6 (units)
   (b) 3 (hundreds of thousands) 2 (hundreds)
   (c) 0 (tens of thousands) 2 (tens)
   (d) 6 (millions) 4 (hundreds of thousands)
2. 270 956

ITEM W – 2   READING LARGE NUMBERS

STEPS

One: *Beginning at the decimal point, separate the digits into groups of three.*

Two: *Read each group of three digits followed by the name of that group.*

NOTE: While recognizing that each individual place has its own value, you will find it helpful to consider the digits in groups of three. Within each group is the same sequence — units, tens, hundreds — as illustrated below.

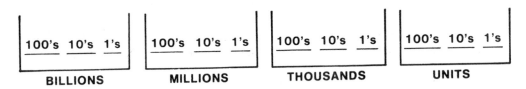

100's 10's 1's	100's 10's 1's	100's 10's 1's	100's 10's 1's
BILLIONS	**MILLIONS**	**THOUSANDS**	**UNITS**

Example A   Read 7046271.

One: 7 046 271 is 7 million, 46 thousand, 271 (units).

Two: seven million, forty-six thousand, two hundred seventy-one (omit the word units).

Example B   Read 684000720.

One: 684 000 720 is 684 million, 0 thousand, 720 (units).

Two: six hundred eighty-four million, seven hundred twenty.
(Since the thousands group is zero, it is omitted in the reading.)

Example C   Read 63480009200.

One: 63 480 009 200 is 63 billion, 480 million, 9 thousand, 200 (units).

Two: sixty-three billion, four hundred eighty million, nine thousand, two hundred.

EXERCISE W – 2

Cover the answer column while you read each of the following numbers. Then check.

3407162	3.407 162	three million, four hundred seven thousand, one hundred sixty-two.
9370	9 370	nine thousand, three hundred seventy.
8000402	8 000 402	eight million, four hundred two.
16027003014	16 027 003 014	sixteen billion, twenty-seven million, three thousand, fourteen.
7007007	7 007 007	seven million, seven thousand, seven.
32000000	32 000 000	thirty-two million.
8004700	8 004 700	eight million, four thousand, seven hundred.

ITEM W – 3 WRITING LARGE NUMBERS

STEPS

One: *Identify the groups (billions, millions, thousands, units).*
Two*: *Prepare a guide with three spaces for each group.*
Three: *Write the numerals in the appropriate group places.*
Four: *Use zeros if necessary so that each group has three places.*

Example A

Write in numerals. Seven billion, sixteen million, thirty thousand, six hundred twenty-three

One: The groups are: billions (7), millions (16), thousands (30), and units (623).

Two: (4 groups; 12 places)

billions	millions	thousands	units

Three: _ _ 7 _ 1 6 _ 3 0 6 2 3

Four: 0 0 7 0 1 6 0 3 0 6 2 3 or 7 016 030 623

Example B

Write in numerals. Eight hundred thirteen million

One: The groups are: millions (813), thousands (0), and units (0)

Two: (3 groups; 9 places)

millions	thousands	units

Three: 8 1 3 _ _ _ _ _ _

Four: 8 1 3 0 0 0 0 0 0 or 813 000 000

Example C

Write in numerals. Twelve billion, eight hundred million, ninety-two

One: 4 groups: billions, millions, thousands, units

Two:

billions	millions	thousands	units

Three: _ 1 2 8 0 0 _ _ _ _ 9 2

Four: 0 1 2 8 0 0 0 0 0 0 9 2 or 12 800 000 092

*This is a rather contrived or artificial approach to writing numbers. It is recommended as a temporary measure only; as soon as you feel confident that you can write numerals correctly, you should dispense with a crutch such as this.

If you went to school a few years ago, you were probably taught to use commas to separate digits into groups of three for convenience in reading and writing large numerals. Some countries use the comma for different purposes and this can lead to confusion. Today, with the growing emphasis on standardization that has accompanied the spread of the metric system, Canada has joined other nations of the world in an effort to standardize the format of written numerals. Under the SI metric guidelines, the use of the comma to separate groups of digits is not acceptable. Instead, an empty space is left which serves the same purpose.

EXAMPLES

Instead of 32,064,277   you would write 32 064 277
Instead of 8,000,650   you would write 8 000 650
If there are only four digits in a numeral, no space is necessary unless the four-digit numeral is used with larger numerals that do require spaces. For example,

4618	4 618
2387	15 600
6598	6 598
No space required	Space is required

EXERCISE W – 3

Write numerals for the following. Cover the answers on the right while you write your own answer. Then compare.

1. Ten million, thirty-five thousand	*10,035,000*	10 035 000
2. Sixteen thousand, two hundred seventeen	*16217*	16 217
3. Four billion, twenty-five million, eight hundred thousand, two hundred seven	*4,025,800,207*	4 025 800 207
4. Six million, six thousand, six	*6,006,006*	6 006 006
5. Five hundred seventy-three million	*570 000 000*	573 000 000
6. Eighty-four million, three hundred forty	*84 000 340*	84 000 340
7. Five hundred sixty-five thousand, fifty	*565 050*	565 050
8. Two thousand twelve	*2012*	2012
9. Three hundred eighty million, four thousand	*380 004 000*	380 004 000

ITEM W – 4 THE FUNDAMENTAL OPERATIONS

The four fundamental operations in arithmetic, whether with whole numbers or fractions or decimals, are: addition, subtraction, multiplication, and division. Two of these are operations that combine; the other two are separating operations.

THE COMBINING OPERATIONS

Addition and multiplication are the combining operations; they bring together two or more quantities and combine them into a single quantity.

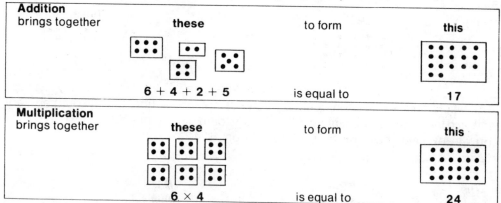

Multiplication may be regarded as a short form of addition in which the quantities to be combined are all alike. For example, 6 multiplied by 4 gives the same result as the addition of four 6s or the addition of six 4s.

$$6 \times 4 \text{ is equal to } 6 + 6 + 6 + 6$$
$$\text{and is also equal to } 4 + 4 + 4 + 4 + 4 + 4$$

Find the result of 7 × 5 and the result of 146 × 4 using addition.

In *addition*, the order of addends does not affect the sum.

EXAMPLE

6 + 4 + 2 + 5 = (6 + 4) + (2 + 5) or 10 + 7	[*Answer:* 17]
or 6 + (4 + 2) + 5 = 6 + 6 + 5	[*Answer:* 17]
or 6 + (4 + 2 + 5) = 6 + 11	[*Answer:* 17]

In *multiplication*, the order of factors does not affect the product.

EXAMPLE

2 × 3 × 8 = (2 × 3) × 8 = 6 × 8 = 48
or 2 × (3 × 8) = 2 × 24 = 48
or (2 × 8) × 3 = 16 × 3 = 48

THE SEPARATING OPERATIONS

Subtraction and division are separating operations in which a single quantity is separated into two parts (subtraction) or into two or more parts (division).

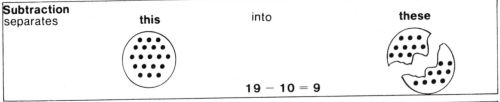

Subtraction separates   **this**   into   **these**

$$19 - 10 = 9$$

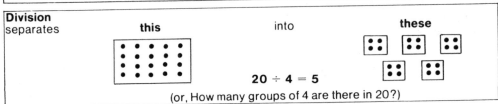

Division separates   **this**   into   **these**

$$20 \div 4 = 5$$

(or, How many groups of 4 are there in 20?)

All whole number division could be done using subtraction, but it would be very slow and laborious. For example, $624 \div 156$ could be interpreted as:
"How many times can 156 be subtracted from 624?"

The	624		*But*		
solution	-156		*this*		
could	468	once	*way*		4
be	-156		*is*	156)	624
like	312	twice	*much*		624
this	-156		*easier*		0
	156	three times			
	-156				
	0	four times			
	Answer: 4			*Answer:* 4	

$20 \div 4$ may have two meanings

(1) How many groups of 4 are there in 20?

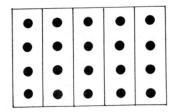

Answer: There are 5 groups.

$$20 \div 4 = 5$$

(2) If 20 is divided into 4 equal parts, how many are there in each part?

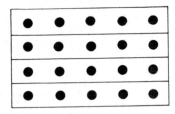

Answer: There are 5 in each part.

$$20 \div 4 = 5$$

ITEM W – 5 FUNDAMENTAL OPERATIONS: TERMINOLOGY

This and the following pages probably contain some words that are unfamiliar to you, although you may have known them at one time. It is not recommended that you try to memorize immediately the material on these pages; rather, be aware that it is available for reference as you come across the terms in subsequent pages.

SUMMARY OF THE FOUR BASIC OPERATIONS

Process	Sign	Meaning	Components or parts	Answer	Inverse operation
Addition	+	Plus	Addends	Sum	Subtraction
Subtraction	−	Minus	Minuend Subtrahend	Difference	Addition
Multiplication	×	Times	*Multiplicand *Multiplier	Product	Division
Division	÷ $\overline{)}$	Divided by Divided into	Dividend Divisor	Quotient	Multiplication

*or factors

EXAMPLES

Addition	
Addend	432
Addend	641
Addend	816
Sum	1889

Subtraction	
Minuend	402
Subtrahend	234
Difference	168

Multiplication	
Multiplicand	275
Multiplier	20
Product	5500

Division may be presented in three different forms.

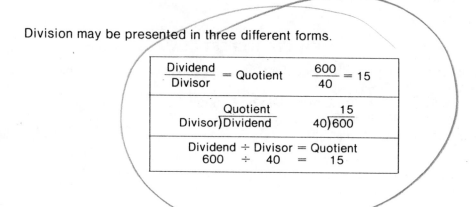

$$\frac{\text{Dividend}}{\text{Divisor}} = \text{Quotient} \qquad \frac{600}{40} = 15$$

$$\frac{\text{Quotient}}{\text{Divisor)}\overline{\text{Dividend}}} \qquad \frac{15}{40)\overline{600}}$$

$$\text{Dividend} \div \text{Divisor} = \text{Quotient}$$
$$600 \div 40 = 15$$

The next few pages define some of the terms that you will meet in the study of whole numbers.

1. Odd and even numbers
2. Prime and composite numbers
3. Factor, common factor, highest common factor
4. Multiple, common multiple, lowest common multiple
5. Average
6. Power, base, exponent
7. Square and square root
8. Inverse operations
9. Order of operations

1. ODD AND EVEN NUMBERS

An odd number is one that ends in 1, 3, 5, 7, or 9.
An odd number will not divide evenly by 2.
Examples of odd numbers: 3, 127, 2445, and 6009.

An even number is one that ends in 2, 4, 6, 8, or 0.
An even number is exactly divisible by 2.
Examples of even numbers: 42, 648, 3000, and 29 946.

2. PRIME AND COMPOSITE NUMBERS

A prime number has only one pair of factors: itself and 1.
3 is a prime number; its only factors are 3×1.
11 is a prime number; its only factors are 11×1.
29 is a prime number; so are 53 and 47 and 61.

A composite number has more than one set of factors.
4 is a composite number; its factors are 4×1 *or* 2×2.
12 is a composite number; $12 = 12 \times 1$ *or* 6×2 *or* 4×3.
25 is composite; its factors are 25×1 *or* 5×5.

3. FACTOR, COMMON FACTOR, HIGHEST COMMON FACTOR

A factor is a number that can be multiplied by another number to form a given product.
3 is a factor of 18; 3 can be multiplied by 6 to get 18.
7 is a factor of 35; 7 multiplied by 5 equals 35.
2 is a factor of any even number.

A factor of a number will divide evenly into that number.
Is 6 a factor of 35? No. 6 will not divide evenly into 35.
Is 3 a factor of 69? Yes. 3 divides evenly into 69.

A common factor of two or more numbers is any number that is a factor of each.
What are the common factors of 18 and 24?
The factors of 18 are 1, 2, 3, 6, 9, and 18
The factors of 24 are 1, 2, 3, 4, 6, 8, 12, and 24.
The common factors of 18 and 24 are: 1, 2, 3, and 6.

The highest common factor of two or more numbers is the largest factor that is common to each.
What is the highest common factor (HCF) of 18 and 24?
The common factors of 18 and 24 are 1, 2, 3, and 6.
The HCF of 18 and 24 is 6 (the largest of the set).

Find the HCF of 45, 60, and 90.
The factors of 45 are 1, 3, 5, 9, 15, and 45.
The factors of 60 are 1, 2, 3, 4, 5, 6, 10, 12, 15, 20, 30, and 60.
The factors of 90 are 1, 2, 3, 5, 6, 9, 10, 15, 18, 30, 45, and 90.
The common factors are 1, 3, 5, and 15.
The HCF is the largest of these, which is 15.

Another way to find the HCF of two or more numbers is to divide by prime factors that are common to each of the numbers.

Example: Find the HCF of 24, 32, and 48.

Solution:

$$
\begin{array}{r|rrr}
2 & 24 & 32 & 48 \\
\hline
2 & 12 & 16 & 24 \\
\hline
2 & 6 & 8 & 12 \\
\hline
 & 3 & 4 & 6 \\
\end{array}
$$

The HCF is the product of all the divisors.
The HCF of 24, 32, and 48 is 2 × 2 × 2, or 8.

---

4. MULTIPLE AND LOWEST COMMON MULTIPLE (LCM)

A multiple of a given number is a product of that number and another number.
45 is a multiple of 3 because 3 can be multiplied by 15 to get 45.
45 is also a multiple of 5, of 9, of 15, and of itself.
To test whether one number is a multiple of another, divide. If the division works out evenly, then the numbers are in a factor-multiple relationship.

EXAMPLE

Is 128 a multiple of 8?   *Test:* 128 ÷ 8 = 16 (works out evenly).
Answer: Yes. 128 is a multiple of 8, and 8 is a factor of 128.

Is 246 a multiple of 16?   *Test:* 246 ÷ 16 = 15 with 6 left over.
Answer: No. 246 is not a multiple of 16, and 16 is not a factor of 246.

The lowest common multiple (LCM) of two or more numbers is the smallest number that is a multiple of each number.

EXAMPLES

What is the LCM of 4 and 5?
Multiples of 4 are 4, 8, 16, 20, 24, 28, etc.
Multiples of 5 are 5, 10, 15, 20, 25, 30, etc.
The lowest multiple that is common to both is 20. The LCM of 4 and 5 is 20.

Find the LCM or 20, 40, and 60.
You can ignore 20 because any multiple of either 40 or 60 has to be a multiple of 20.
The multiples of 40 are 40, 80, 120, 160, etc.
The multiples of 60 are 60, 120, 180, etc.
The LCM of 20, 40, and 60 is 120.

What is the LCM of 12, 16, and 18?
Select the largest number, 18.
Write the multiples of 18 until you reach one that is also a multiple of 12 and 16.
The multiples of 18 are 18, 36, 54, 72, 90, 108, 126, 144, etc.
144 is the first number that is also a multiple of 12 and 16.
The LCM of 12, 16, and 18 is 144.

5. AVERAGE

The average is the total divided by the number of parts. To find the average, first find the sum, then divide by the number of addends.
What is the average of 22, 17, 31, and 14?
The sum of the numbers is 84.
There are four addends.
$84 \div 4 = 21$
The average of 22, 17, 31, and 14 is 21.

Find the average of $1.25, $6.00, and $2.50.

The average is $\dfrac{\$1.25 + \$6.00 + \$2.50}{3} = \dfrac{\$9.75}{3} = \$3.25.$

Sometimes the total is given, eliminating the necessity of addition.

223 people contributed a total of $1928.95 to the cancer fund.
What was the average contribution?

The average contribution was $\dfrac{\$1928.95}{223}$ or $8.65.

6. POWER, BASE, AND EXPONENT

A power is the result of multiplying a number by itself a specified number of times.
In the power 10^4, 10 is multiplied by itself 4 times.
10^4 means $10 \times 10 \times 10 \times 10$
10^4 is the fourth power of 10; it is equal to 10 000
A power is composed of a *base* and an *exponent*.

The base is the number that is multiplied. In the power 10^4, the base is 10.

The exponent tells the number of times that the base is multiplied.
In the power 10^4, the exponent is 4.
The exponent 4 tells that 10 is multiplied by itself four times.

EXAMPLE

5^3 is the third power of 5.
It is read "5 to the third" or "5 cubed."
The base is 5; the exponent is 3.

$5^3 = 5 \times 5 \times 5$ or 125

EXAMPLE

2^5 is the fifth power of 2.
It is read "2 to the fifth."
The base is 2; the exponent is 5.

$2^5 = 2 \times 2 \times 2 \times 2 \times 2$ or 32

7. SQUARE AND SQUARE ROOT

The square of a number is the result obtained when that number is multiplied by itself.
The square of 8 is 64, or $8^2 = 64$
The square of 15 is 225, or $15^2 = 225$

The square root of a number is one of its two equal factors.

The square root of 36 is 6.
The square root of 625 is 25.
The square root of 144 is 12.

The symbol for square root is $\sqrt{}$

$\sqrt{36} = 6$ $\sqrt{81} = 9$ $\sqrt{1600} = 40$

The squares of the numbers from 1 to 20 are listed below:

$1^2 = 1$	$6^2 = 36$	$11^2 = 121$	$16^2 = 256$
$2^2 = 4$	$7^2 = 49$	$12^2 = 144$	$17^2 = 289$
$3^2 = 9$	$8^2 = 64$	$13^2 = 169$	$18^2 = 324$
$4^2 = 16$	$9^2 = 81$	$14^2 = 196$	$19^2 = 361$
$5^2 = 25$	$10^2 = 100$	$15^2 = 225$	$20^2 = 400$

And it is useful to know that

$25^2 = 625$ $30^2 = 900$ $40^2 = 1600$ $50^2 = 2500$

Methods of finding the square root are so varied that they will not be examined at this point.

8. INVERSE OPERATIONS

An inverse operation is one that nullifies, voids, or cancels out another operation. If you turn on a light switch, then turn it off, the second operation undoes the first.

Addition and Subtraction are inverse operations.

If you begin with 28, then add 6, then subtract 6, the second operation of subtraction undoes the first operation of addition.

$$28 + 6 - 6 = 28$$

Similarly, $52 - 9 + 9 = 52$ (Here addition undoes subtraction.)

Multiplication and Division are inverse operations.

If you start with 70, multiply it by 15, then divide the result by 15, the operation of division undoes the previous operation of multiplication.

$$70 \times 15 \div 15 = 70$$

Similarly, $63 \div 21 \times 21 = 63$ (Here multiplication undoes division.)

An understanding of inverse operations is useful in proving answers, in rearranging formulas, and in solving equations.

9. ORDER OF OPERATIONS

If a problem in arithmetic or algebra involves two or more of the operations of addition, subtraction, multiplication or division, these operations must be performed in accordance with an established sequence. Basically, the sequence (or order) is that multiplication and division be done before addition or subtraction. If, however, any numbers are enclosed in brackets, these numbers must be "done" first regardless of the sign of operation within the bracket. The order, then, is:

First: Brackets
Second: Multiplication and Division (in order of occurrence)
Third: Addition and Subtraction (in order of occurrence)

EXAMPLES

1. Calculate the following: $2 + 8 \times 12 - 6 \div 2$

   Multiply 8×12 $2 + \quad 96 \quad - 6 \div 2$
   Divide $6 \div 2$ $2 + \quad 96 \quad - \quad 3$
   Add $2 + 96$ $98 \quad - \quad 3$
   Subtract $98 - 3$ 95 Answer is 95

2. Here are the same numbers but with brackets.

   Calculate the following: $(2 + 8) \times 12 - 6 \div 2$
   Brackets $2 + 8 = 10$ $10 \quad \times 12 - 6 \div 2$
   Multiply 10×12 $120 \quad - 6 \div 2$
   Divide $6 \div 2$ $120 \quad - \quad 3$
   Subtract $120 - 3$ 117 Answer is 117

3. And the same numbers but with a different set of brackets.

   Find the answer: $2 + 8 \times (12 - 6 \div 2)$
   Brackets $6 \div 2$ $2 + 8 \times (12 - \quad 3 \quad)$
   Brackets $12 - 3$ $2 + 8 \times \quad 9$
   Multiply 8×9 $2 + \quad 72$
   Add $2 + 72$ 74 Answer is 74

Try these. The answers are included.

(a) $3(5 + 9) - 2(15 - 3) = 18$

(b) $\dfrac{64 \div 2 + 7}{33 - 4 \times 5} = 3$

(c) $2(3^2 - 7) + 18 - 8 = 14$

(d) $(3 + 4) \times 6 - 5 \times 3 = 27$

(e) $\dfrac{7 \times 4 - (13 + 6)}{3 \times 4 - 3} = 1$

(f) $\dfrac{4 \times 2^3}{2} - 3(6 - 2) = 4$

(g) $(5 + 2 \times 3)(12 - 6 \div 3) = 110$

(h) $15 + \dfrac{1}{2}$ of $20 \div 2 = 20$

EXERCISE W – 5

1. Which of the following are even numbers? 62, 347, 1090, 7, 836, 775.
2. Which of these numbers are prime? 81, 67, 9, 2, 31, 49.
3. What is the only even prime number?
4. Give one factor (other than the number itself or 1) to prove that each of the following is a composite number.
   (a) 91   (b) 87   (c) 55   (d) 65   (e) 39
5. What are the common factors of 24 and 42?
6. What is the HCF of 24 and 42?
7. Name the first three multiples of 27.
8. Which of the following are multiples of 15?
   25, 30, 40, 55, 95, 120.
9. Find the average of each set:
   (a) 25, 56, 39, 17, and 23.
   (b) 178, 200, and 162.
10. Write the square of each of the following:
    (a) 12   (b) 9   (c) 100
11. Write the power that has a base of 10 and an exponent of 3.
12. Write a power composed of base 2 and exponent 6.
13. Find the value of each of the following:
    (a) 4^3 (b) 10^5 (c) 3^2 (d) 7^2 (e) 1^4
14. Calculate the following in accordance with the correct order of operations.
    (a) $8 - 3 \times 2$ (b) $6 \times 3 - 8 \div 4$
    (c) $2 - (7 - 6) + 4$ (d) $12 + 9 \div 3 - 2 \times 4$
    (e) $2 + 4 \times 15 \div 5$ (f) $8^2 - 3(6 - 2)^2$
    (g) $8 \times 9 - 6 \times 4 + \dfrac{36}{6}$ (h) $(4 \times 6)(12 - 5)$

ANSWERS FOR EXERCISE W – 5

1. 62, 1090, 836   2. 67, 2, 31   3. 2
4. (a) $91 = 13 \times 7$ (b) $87 = 3 \times 29$ (c) for 55 could be 11 or 5
   (d) for 65 could be 13 or 5   (e) for 39 could be 13 or 3
5. The common factors of 24 and 42 are 1, 2, 3, 6.
6. The HCF is 6.
7. 27, 54, 81   8. 30, 120   9. (a) 32   (b) 180
10. (a) $12^2 = 144$ (b) $9^2 = 81$ (c) $100^2 = 10\,000$
11. 10^3 12. 2^6
13. (a) 64   (b) 100 000   (c) 9   (d) 49   (e) 1
14. (a) 2   (b) 16   (c) 5   (d) 7
    (e) 14   (f) 16   (g) 54   (h) 168

ITEM W–6(I) COMMON ERRORS IN ADDITION

1. ASSUMING THAT AN ANSWER IS CORRECT WITHOUT CHECKING

Since the process of addition seems relatively easy, some students don't bother to check the answer. If you have a tendency to be inaccurate (the test at the end of the chapter will provide some indication), you should do each question twice, once adding from the top and once adding from the bottom.

EXAMPLE

```
Add
    3
4 5 7      Thinking process for units column:
6 3 8          From the top   7, 15, 19, 28, 31
9 2 4          From the bottom   3, 12, 16, 24, 31
5 6 9
8 9 3
───
    1
```

2. CARELESS ARRANGEMENT OF COLUMNS

```
  4 1 6 2 3            4 1   6 2 3
    74 81              7   4 8 1
        9 9                    9 9
    2   3 0 5          2   3 0 5
61 12 1   4          6 1 1   2 1 4
─────────            ─────────
  Example A            Example B
```

Example A, shown above in a somewhat exaggerated form, will result in an incorrect answer more readily than will Example B. If you have trouble keeping columns in line, you might try using graph paper. It is customary in writing large numbers with 5 or more digits to leave a space between each group of three digits, as shown in Example B.

3. INCORRECT CARRYING

Most students realize that if the sum of a column is a two-digit number, the tens digit is carried to the next column. Occasionally, instead of doing this, a person will write both digits in the answer space.

EXAMPLE

```
    4 6 3              4 6 3
    7 8 2              7 8 2
    6 5 1              6 5 1
    9 4 4              9 4 4
   ─────              ─────
   262310             2 8 4 0
   Incorrect          Correct
```

ITEM W – 6(II)   COMMON ERRORS IN SUBTRACTION

1. FAILURE TO RE-GROUP (OR BORROW) CORRECTLY

There are two methods used in the subtraction of whole numbers. The method that you learned is probably a part of your subconscious, and you would have considerable difficulty if you were to try to adopt the other method. The thinking processes for both are shown here so that you can identify and relate to the one that you have learned.

EXAMPLE A	Thinking process Method One	Same example	Thinking process Method Two
6 0 0 4 0	6 from 10 is 4	6 0 0 4 0	6 from 10 is 4
3 4 1 3 6	4 from 4 is 0	3 4 1 3 6	3 from 3 is 0
2 5 9 0 4	1 from 10 is 9	2 5 9 0 4	1 from 10 is 9
	5 from 10 is 5		4 from 9 is 5
	4 from 6 is 2		3 from 5 is 2

EXAMPLE B	Method One	Same example	Method Two
9 0 0 0	8 from 10 = 2	9 0 0 0	8 from 10 = 2
2 4 0 8	1 from 10 = 9	2 4 0 8	0 from 9 = 9
6 5 9 2	5 from 10 = 5	6 5 9 2	4 from 9 = 5
	3 from 9 = 6		2 from 8 = 6

2. SUBTRACTING A SMALLER DIGIT FROM A LARGER REGARDLESS OF WHICH IS MINUEND AND WHICH IS SUBTRAHEND

The examples below are not frequent errors but neither are they rare.

6 2 0 3 4	5 9 7 2 8		6 2 0 3 4	5 9 7 2 8
1 6 8 9 2	4 2 9 8 1		1 6 8 9 2	4 2 9 8 1
5 4 8 6 2	1 7 2 6 7		4 5 1 4 2	1 6 7 4 7
These are incorrect			These are correct	

PRACTICE IN SUBTRACTION

Cover the answers with a card or a piece of paper while you write your own answers. Then compare.

60 042	7321	96 000	68 045	7316	4699
18 659	6804	26 040	29 999	2084	3149
41 383	517	69 960	38 046	5232	1550

12 000	82 134	2417	8694	15 515	62 300
6 054	69 875	1000	2179	86	52 301
5 946	12 259	1417	6515	15 429	9 999

ITEM W–6(III) COMMON ERRORS IN MULTIPLICATION

1. INCORRECT PLACEMENT OF PARTIAL PRODUCTS

This is incorrect.

```
    4 3 1 6
  × 2 7 5
  2 1 5 8 0
  3 0 2 1 2
    8 6 3 2
  6 0 4 2 4
```

These are correct.

```
      4 3 1 6        4 3 1 6          4 3 1 6
    × 2 7 5         × 2 7 5          × 2 7 5
    2 1 5 8 0     8 6 3 2 0 0        2 1 5 8 0
  3 0 2 1 2       3 0 2 1 2 0      3 0 2 1 2 0
  8 6 3 2           2 1 5 8 0      8 6 3 2 0 0
  1 1 8 6 9 0 0   1 1 8 6 9 0 0    1 1 8 6 9 0 0
```

2. FAILURE TO INDENT PROPERLY WHEN THE MULTIPLIER ENDS WITH 0

This is incorrect.

```
    4 0 6 2 3
    × 7 2 0
    8 1 2 4 6 0
  2 8 4 3 6 1
  3 6 5 6 0 7 0
```

These are correct.

```
      4 0 6 2 3            4 0 6 2 3
      × 7 2 0              × 7 2 0
      8 1 2 4 6 0          0 0 0 0 0
    2 8 4 3 6 1            8 1 2 4 6
    2 9 2 4 8 5 6 0      2 8 4 3 6 1
                          2 9 2 4 8 5 6 0
```

3. INCORRECT PLACEMENT WHEN SEVERAL ZEROS ARE INVOLVED

This is incorrect.
2 0 0 3
× 6 0 4 0
8 0 1 2 0
1 2 0 1 8 0
1 2 8 1 9 2 0

These are correct.	
2 0 0 3	2 0 0 3
× 6 0 4 0	× 6 0 4 0
8 0 1 2 0	0 0 0 0
1 2 0 1 8 0	8 0 1 2
1 2 0 9 8 1 2 0	0 0 0 0
	1 2 0 1 8
	1 2 0 9 8 1 2 0

4. CONFUSING MULTIPLICATION BY 0 WITH MULTIPLICATION BY 1

This is a very common error.

When any number is multiplied by 0, the answer is 0.

When any number is multiplied by 1, the answer is the same as the number that was multiplied.

$$7 \times 0 = 0 \text{ but } 7 \times 1 = 7$$
$$0 \times 9 = 0 \text{ but } 1 \times 9 = 9$$

This is incorrect.
2 3 6 7
4 0 2
4 7 3 4
2 3 6 7
9 4 6 8
9 7 5 2 0 4

This is correct.
2 3 6 7
4 0 2
4 7 3 4
0 0 0 0
9 4 6 8
9 5 1 5 3 4

PRACTICE EXERCISE

Cover answers while you work out your own. Then compare.

740	8006	9692	7640	5972
620	408	89	3004	6050
14800	64048	87228	30560	298600
4440	320240	77536	2292000	358320
458800	3266448	862588	22950560	36130600

ITEM W–6(IV) COMMON ERRORS IN DIVISION

1. OMISSION OF NECESSARY ZEROS IN THE QUOTIENT

This is incorrect.	This is correct.
3 0 4 5) 1 5 0 2 0	3 0 0 4 5) 1 5 0 2 0

This error can be avoided by placing the first digit in the quotient exactly above the corresponding digit in the dividend, and then making sure that each following digit in the dividend has an answering digit above it.

Study the difference in the two examples above, as well as in the following:

```
      2 0 3              2 0 0 3
31) 6 2 0 9 3      31) 6 2 0 9 3
    6 2                  6 2
    0 0 9 3              0 0 9 3
        9 3                  9 3
         0                   0
```
Incorrect	Correct

```
      2 4                 2 4 0
83) 1 9 9 2 0      83) 1 9 9 2 0
    1 6 6              1 6 6
    3 3 2              3 3 2
    3 3 2              3 3 2
     0 0               0 0
```

2. SUPERFLUOUS ZEROS IN THE QUOTIENT

```
  204010*               2 4 1
25) 6 0 4 3      25) 6 0 4 3
    5 0               5 0
    1 0 4             1 0 4
    1 0 0             1 0 0
        4 3             4 3
        2 5             2 5
        1 8             1 8
```

*This error is the result of the wrong sequence of steps in division. Always bring down the next number *before* you say how many times the divisor will go into the dividend.

STEPS IN DIVISION
(a) How many times?
(b) Multiply
(c) Subtract
(d) Bring down
Repeat this sequence

PRACTICE IN DIVISION

Work the following. Then compare your work with the solutions on this page.

(a) 27060 ÷ 3	(b) 4020 ÷ 5	(c) 138069 ÷ 23
(d) 24840 ÷ 12	(e) 1550062 ÷ 31	(f) 14880 ÷ 62
(g) 214375 ÷ 343	(h) 143027 ÷ 198	(i) 257441 ÷ 642
(j) 257441 ÷ 250	(k) 191100 ÷ 250	(l) 403021 ÷ 467

Solutions

(a)
```
       9020
   3) 27060
```

(b)
```
        804
   5) 4020
```

(c)
```
        6003
  23) 138069
      138
      0069
        69
         0
```

(d)
```
       2070
  12) 24840
      24
       084
        84
        00
```

(e)
```
       50002
  31) 1550062
      155
       00062
          62
           0
```

(f)
```
        240
  62) 14880
      124
       248
       248
        00
```

(g)
```
        625
 343) 214375
      2058
       857
       686
      1715
      1715
         0
```

(h)
```
        722
 198) 143027
      1386
       442
       396
       467
       396
        71
```

(i)
```
        400
 642) 257441
      2568
      0641
```

(j)
```
       1029
 250) 257441
      250
       744
       500
      2441
      2250
       191
```

(k)
```
        764
 250) 191100
      1750
      1610
      1500
      1100
      1000
       100
```

(l)
```
        863
 467) 403021
      3736
      2942
      2802
      1401
      1401
         0
```

ITEM W-7 PROVING AN ANSWER

To prove an answer in any one of the four operations you may use the inverse operation.

1. ADDITION IS PROVED BY SUBTRACTION

To prove an answer in addition, subtract the addend from the sum. The result is equal to the other addend.

Example of addition		*Proof by subtraction*	
Addend	4 2 6	Sum	1 0 6 5
+ Addend	6 3 9	− Addend	4 2 6
Sum	1 0 6 5	Other addend	6 3 9

2. SUBTRACTION IS PROVED BY ADDITION

To prove an answer in subtraction, add the difference and the subtrahend. The result is equal to the minuend.

Example of subtraction		*Proof by addition*	
Minuend	6 0 0 2	Subtrahend	1 7 8 8
− Subtrahend	1 7 8 8	+ Difference	4 2 1 4
Difference	4 2 1 4	Minuend	6 0 0 2

3. MULTIPLICATION IS PROVED BY DIVISION

To prove an answer in multiplication, divide the product by one of the factors. The result is equal to the other factor.

Example of multiplication

Factor	6 2 0
Factor	7
Product	4 3 4 0

Proof by division

$$\frac{\text{Product}}{\text{Factor}} \quad \frac{4340}{7} = 620 \quad \text{or} \quad \frac{4340}{620} = 7$$

4. DIVISION IS PROVED BY MULTIPLICATION

To prove an answer in division, multiply the quotient by the divisor, and add the remainder if there is one. The result is equal to the dividend.

Example of division

```
              27  Quotient
Divisor  23) 650  Dividend
             46
            190
            161
             29
```

Proof by multiplication

Quotient	27
Divisor	× 23
	81
	54
	621
Remainder	+ 29
Dividend	650

PRACTICE WITH WHOLE NUMBERS

1. Multiplication

(a) $537\,084 \times 69$ (b) 390×460 (c) $7 \times 21 \times 40$

(d) 4068×200 (e) $25\,975 \times 84$ (f) 500×300

(g) 8050×2060 (h) $9 \times 65 \times 10$ (i) $100\,069 \times 79$

(j) 634×209 (k) 3700×470 (l) 325×548

2. Division

(a) $639\,865 \div 56$ (b) $40\,000 \div 24$ (c) $2498 \div 300$

(d) $2189 \div 70$ (e) $42\overline{)840\,231}$ (f) $20\overline{)450\,286}$

(g) $95\overline{)764\,035}$ (h) $381\overline{)648\,829}$ (i) $\dfrac{72\,973}{800}$

(j) $\dfrac{35\,791}{29}$ (k) $\dfrac{200\,002}{17}$ (l) $\dfrac{1749}{1430}$

3. Powers

Which is larger and by how much?

(a) 9^2 or 3^5 (b) 18^2 or 7^3 (c) 14^2 or 6^3

(d) 1^4 or 2^2 (e) 3^5 or 15^2 (f) 2^5 or 6^2

4. Order of operations

Find the result of each of the following.

(a) $20 - (3 + 5 \times 2)$ (b) $18 \div 6 + 4 \div 2 \times 3$

(c) $(19 + 5) \div 4 \div (3 \times 2)$ (d) $5(12 - 3 \times 3) + 2 \times 3^2$

(e) $40 - (24 \div 4 + 8)$ (f) $2(3^2 - 7) + 16 - 4$

5. Place value

(a) The number 16 240 975 has ____ in the hundreds place and ____ in the millions place.

(b) Write the number that has 7 tens of thousands, 6 units, 4 millions, 5 hundreds of thousands, 2 tens, 0 thousands, and 9 hundreds.

(c) Write in numerals: forty million, eight thousand, six hundred ten.

(d) What is the place value of the digit 3 in each of the following?

    (i) 243 600 298   (ii) 75 835 201

    (iii) 45 963   (iv) 31 456 729 000

6. Miscellaneous terminology

(a) The numbers are 480 and 12. The product is _____, the sum is _____, the quotient is _____, and the difference is _____.

(b) Copy only the prime numbers: 41, 63, 85, 87, 89, 17, 101, 23

(c) Is 30 a factor of 10 or a multiple of 10?

(d) Write all the factors of 48.

(e) Write the first three multiples of 24.

(f) Find the average of 6, 18, 7, and 13.

(g) Fill in the missing words. Addition and _____ are inverse operations. Multiplication is the inverse of _____ .
The two operations that combine are _____ and _____ .

Answers are on page 32.

WHOLE NUMBER TEST A
(see page 118.)

Addition

			3	66	86	26 074
		372	8	72	83	9 271
28	342	964	9	803	74	86 458
216	464	4633	6	79	32	42
189	94	776	2039	5	2416	97	97
+265	+987	+867	+676	+4	+489	+63	+80 269

Subtraction

73	900	63 000	9996	23 157	500	8003	62 030
−32	−565	−27 856	−4998	−7 048	−399	−2694	−17 694

Multiplication

573	2436	258 634	964	6995	3527	60 802	960
×6	×45	×78	×736	×284	×925	×7005	×760

Division

Write your answers with quotient and remainder; you do not need to work beyond the decimal point. For example,

$$423 \div 5 = 84 \text{ R } 3$$

(a) 296 ÷ 4   (b) 694 ÷ 32   (c) 428 ÷ 123

(d) 9 900 066 ÷ 33   (e) 80 465 ÷ 47   (f) 691 027 ÷ 523

(g) 72 958 ÷ 73   (h) 800 479 ÷ 351

WHOLE NUMBER TEST B
(see page 118.)

Addition

				2	56	46	16 004
			372	8	72	83	9 271
	25	742	924	4	903	75	86 358
	116	463	1633	6	75	22	42
139	34	976	1059	9	1816	96	27
+275	+987	+857	+876	+3	+469	+53	+10 069

Subtraction

64	2005	42 000	9995	12 157	600	7002	92 040
−32	−1639	−16 894	−8997	−6 148	−299	−1693	−15 690

Multiplication

462	4263	825 634	694	5996	5372	80 602	690
×7	×45	×87	×673	×248	×259	×5007	×670

Division

(a) $296 \div 3$

(b) $252 \div 12$

(c) $426 \div 124$

(d) $8\,400\,042 \div 21$

(e) $90\,456 \div 57$

(f) $691\,072 \div 423$

(g) $62\,958 \div 63$

(h) $800\,479 \div 251$

WHOLE NUMBER TEST C
(see page 118.)

Addition

				2	54	48	16 005
			874	6	72	83	7 293
	27	497	372	4	909	75	86 358
	141	364	922	5	75	22	44
136	36	677	1333	9	1816	98	27
+275	+985	+558	+1055	+3	+468	+63	+30 069

Subtraction

64	2004	9992	61 000	12 157	600	7002	92 040
−32	−1537	−8997	−16 873	−6 344	−289	−1973	−17 490

Multiplication

624	4463	528 634	703	4792	6376	80 406	830
×9	×47	×78	×677	×48	×341	×5009	×890

Division

(a) $327 \div 3$ (b) $264 \div 12$ (c) $384 \div 132$

(d) $5\,100\,085 \div 17$ (e) $87\,493 \div 57$ (f) $786\,402 \div 512$

(g) $71\,992 \div 72$ (h) $600\,342 \div 263$

For additional practice with whole numbers see pages 203 to 208.

ANSWERS FOR PRACTICE WITH WHOLE NUMBERS (from page 28.)

1. MULTIPLICATION

(a)	37 058 796	(b)	179 400	(c)	5880	(d)	813 600
(e)	2 181 900	(f)	150 000	(g)	16 583 000	(h)	5850
(i)	7 905 451	(j)	132 506	(k)	1 739 000	(l)	178 100

2. DIVISION

(a)	11 426 R 9	(b)	1666 R 16	(c)	8 R 98	(d)	31 R 19
(e)	20 005 R 21	(f)	22 514 R 6	(g)	8042 R 45	(h)	1702 R 367
(i)	91 R 173	(j)	1234 R 5	(k)	11 764 R 14	(l)	1 R 319

3. POWERS

   (a) 3^5 is larger by 162 $(243 - 81)$ (b) 7^3 is larger by 19 $(343 - 324)$

   (c) 6^3 is larger by 20 $(216 - 196)$ (d) 2^2 is larger by 3 $(4 - 1)$

   (e) 3^5 is larger by 18 $(243 - 225)$ (f) 6^2 is larger by 4 $(36 - 32)$

4. ORDER OF OPERATIONS

   (a) 7   (b) 9   (c) 1   (d) 33   (e) 26   (f) 16

5. PLACE VALUE

   (a) 9 in the hundreds and 6 in the millions.

   (b) 4 570 926   (c) 40 008 610

   (d) (i) millions   (ii) tens of thousands   (iii) units   (iv) tens of billions

6. MISCELLANEOUS TERMINOLOGY

   (a) Product is 5760, sum is 492, quotient is 40, difference is 468.

   (b) 41, 89, 17, 101, 23.   (c) Multiple

   (d) 1, 2, 3, 4, 6, 8, 12, 16, 24, 48.   (e) 24, 48, 72.

   (f) 11   (g) subtraction; division; addition and multiplication.

CHAPTER TWO: **FRACTIONS**

The word "fraction" comes from a Latin word that means "to break." A fraction is sometimes loosely defined as a part of a whole. More precisely, it is "one or more of the equal parts into which a whole is divided."

There are two types of fractions, *common* and *decimal*. Usually, when the word "fraction" is used alone, it refers to a common fraction. Decimal fractions are more often called decimals.

Opinions differ as to the relevance of common fractions in the modern mathematical world. Some say that fractions have become obsolete for two reasons:

1. The gradual replacement of the British system of weights and measures with the metric system

2. The relative ease of computation with decimal fractions as opposed to common fractions, particularly with the widespread use of the hand calculator.

Others feel that an awareness of the concepts involved in common fractions is vital to the understanding of the number system as a whole.

At the present time, standardized tests in mathematics, as well as most of the locally prepared admission tests, do require a knowledge of common fractions.

ITEMS IN THIS CHAPTER

ITEM F-1 MEANING AND TERMINOLOGY

1. NUMERATOR AND DENOMINATOR

A common fraction has two written parts in the form $\frac{a}{b}$.

The upper part is the numerator; the lower part is the denominator. As you examine the diagrams below you will see that in each case:

- a whole has been divided into a number of equal parts
- the denominator tells the total number of equal parts
- the numerator tells the specific number of equal parts

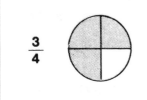

There are 4 equal parts
Each part is one fourth
The whole is equal to $\frac{4}{4}$
The shaded part is $\frac{3}{4}$

There are 6 equal parts
Each part is one sixth
The whole is equal to $\frac{6}{6}$
The shaded part is $\frac{1}{6}$

There are 9 equal parts
Each part is one ninth
The whole is equal to $\frac{9}{9}$
The shaded part is $\frac{5}{9}$

2. A FRACTION EXPRESSES DIVISION

This is another meaning of a fraction in which $\dfrac{\text{Numerator}}{\text{Denominator}} = \dfrac{\text{Dividend}}{\text{Divisor}}$

$\frac{20}{4}$ means 20 divided by 4	$\frac{16}{5}$ means 16 divided by 5	$\frac{3}{4}$ means "3 divided by 4"
$\frac{20}{4} = 20 \div 4 = 5$	$\frac{16}{5} = 16 \div 5 = 3\frac{1}{5}$	$\frac{3}{4} = 3 \div 4 = \frac{3}{4}$ or 0.75

3. PROPER AND IMPROPER FRACTIONS

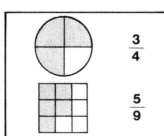

$\frac{3}{4}$

$\frac{5}{9}$

A common fraction may be proper or improper.

A proper fraction is less than one whole.

An improper fraction is equal to or greater than one whole.

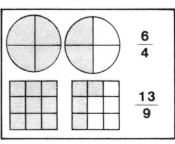

$\frac{6}{4}$

$\frac{13}{9}$

Proper Fractions   **Improper Fractions**

4. MIXED NUMBER

A mixed number is a combination of a whole number and a fraction. An improper fraction may be changed to a mixed number by dividing the denominator into the numerator.

$\dfrac{4}{4} = 4\overline{)4} = 1$	$\dfrac{10}{6} = 6\overline{)10} = 1\dfrac{4}{6}$	$\dfrac{14}{9} = 9\overline{)14} = 1\dfrac{5}{9}$

To change a mixed number to an improper fraction, the procedure is reversed. You multiply the whole number by the denominator, add the numerator, and this amount is placed over the denominator.

$$\frac{\textbf{Whole number} \times \textbf{Denominator} + \textbf{Numerator}}{\textbf{Denominator}}$$

$6\dfrac{2}{3} = \dfrac{6 \times 3 + 2}{3} = \dfrac{20}{3}$	$8\dfrac{1}{5} = \dfrac{8 \times 5 + 1}{5} = \dfrac{41}{5}$	$2\dfrac{5}{9} = \dfrac{2 \times 9 + 5}{9} = \dfrac{23}{9}$

5. EQUIVALENT FRACTIONS

Equivalent fractions are equal in value but expressed in different terms. The diagram below shows an example of equivalence.

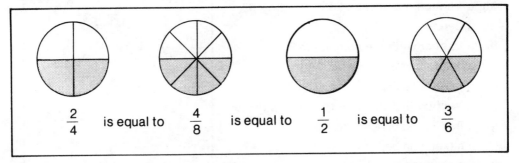

$\dfrac{2}{4}$ is equal to $\dfrac{4}{8}$ is equal to $\dfrac{1}{2}$ is equal to $\dfrac{3}{6}$

A fraction may be changed to an equivalent by multiplying or dividing each part by the same number. If you multiply each part, you raise the fraction to higher terms; if you divide each part, you reduce the fraction to lower terms.

$$\frac{12}{16} \text{ raised to higher terms} = \frac{12 \times 3}{16 \times 3} \text{ or } \frac{36}{48}$$

$$\frac{12}{16} \text{ reduced to lower terms} = \frac{12 \div 4}{16 \div 4} \text{ or } \frac{3}{4}$$

The fractions $\dfrac{12}{16}$, $\dfrac{36}{48}$, and $\dfrac{3}{4}$ are equivalent.

6. UNIT FRACTION

A unit fraction has a numerator of 1. Some examples of unit fractions are $\frac{1}{4}$, $\frac{1}{10}$, $\frac{1}{25}$, and $\frac{1}{8}$.

7. RECIPROCAL

The reciprocal of a given number is the result of dividing 1 by that number.

EXAMPLES

The reciprocal of 4 is $\frac{1}{4}$ The reciprocal of 100 is $\frac{1}{100}$

The reciprocal of $\frac{5}{9}$ is $\frac{9}{5}$ The reciprocal of $\frac{7}{16}$ is $\frac{16}{7}$

The reciprocal of $4\frac{1}{4}$ is $\frac{4}{17}$ The reciprocal of $11\frac{1}{2}$ is $\frac{2}{23}$

EXERCISE F – 1

1. From the numbers at the right, select an example of each of the following:
   - (a) an improper fraction
   - (b) a decimal fraction
   - (c) a unit fraction
   - (d) a fraction that means "4 divided by 9"
   - (e) a percent
   - (f) a fraction with the numerator 3
   - (g) a whole number
   - (h) a proper fraction
   - (i) a fraction that is equal to a whole number
   - (j) a fraction equivalent to one-half
   - (k) a mixed number
   - (l) the reciprocal of $3\frac{1}{3}$

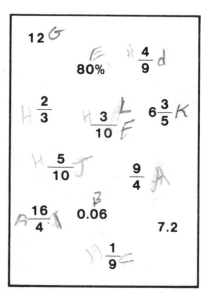

2. Change the following mixed numbers to improper fractions.

    (a) $6\frac{1}{2}$ (b) $2\frac{2}{3}$ (c) $1\frac{1}{4}$ (d) $9\frac{5}{8}$

    (e) $11\frac{1}{10}$ (f) $5\frac{7}{9}$ (g) $2\frac{5}{12}$ (h) $20\frac{4}{5}$

3. Change each of the following to a whole or a mixed number.

    (a) $\frac{11}{4}$ (b) $\frac{12}{3}$ (c) $\frac{16}{5}$ (d) $\frac{8}{1}$

    (e) $\frac{9}{2}$ (f) $\frac{121}{4}$ (g) $\frac{65}{6}$ (h) $\frac{7}{6}$

ANSWERS FOR EXERCISE F – 1

1. (a) $\frac{9}{4}$ or $\frac{16}{4}$ (b) 0.06   (c) $\frac{1}{9}$ (d) $\frac{4}{9}$ (e) 80%

    (f) $\frac{3}{10}$ (or $\frac{3}{5}$)   (g) 12 (or 6 or 7)   (h) $\frac{2}{3}$ or $\frac{3}{10}$ or $\frac{4}{9}$ or $\frac{5}{10}$ or $\frac{1}{9}$

    (i) $\frac{16}{4}$ (equal to 4)   (j) $\frac{5}{10}$ (k) $6\frac{3}{5}$ or 7.2   (l) $\frac{3}{10}$

2. (a) $\dfrac{2 \times 6 + 1}{2} = \dfrac{13}{2}$ (b) $\dfrac{3 \times 2 + 2}{3} = \dfrac{8}{3}$ (c) $\dfrac{4 \times 1 + 1}{4} = \dfrac{5}{4}$

    (d) $\dfrac{8 \times 9 + 5}{8} = \dfrac{77}{8}$ (e) $\dfrac{111}{10}$ (f) $\dfrac{52}{9}$ (g) $\dfrac{29}{12}$ (h) $\dfrac{104}{5}$

3. (a) $2\frac{3}{4}$ (b) 4   (c) $3\frac{1}{5}$ (d) 8

    (e) $4\frac{1}{2}$ (f) $30\frac{1}{4}$ (g) $10\frac{5}{6}$ (h) $1\frac{1}{6}$

ITEM F–2(I) EQUIVALENT FRACTIONS: RAISING TO HIGHER TERMS

STEPS IN RAISING TO HIGHER TERMS

One: Select the number that you will use as a multiplier.
Two: Multiply numerator and denominator by that number.

Example A

Write three fractions equivalent to $\frac{3}{5}$. $\frac{6}{10}$ $\frac{9}{15}$ $\frac{12}{20}$

STEPS

One: Suppose the numbers you select as multipliers are: (a) 7 (b) 2 (c) 12
(You could choose any other numbers.)

Two: (a) $\dfrac{3 \times 7 = 21}{5 \times 7 = 35}$ (b) $\dfrac{3 \times 2 = 6}{5 \times 2 = 10}$ (c) $\dfrac{3 \times 12 = 36}{5 \times 12 = 60}$

$$\frac{3}{5} = \frac{21}{35} = \frac{6}{10} = \frac{36}{60}$$

Sometimes the multiplier is predetermined, as shown in the next example.

Example B

Complete the following. (a) $\dfrac{4}{5} = \dfrac{12}{?15}$ (b) $\dfrac{7}{16} = \dfrac{?}{32}$ (c) $\dfrac{5}{8} = \dfrac{?}{40}$

STEPS

One: In (a) the multiplier has to be 3   (4 × 3 = 12)
In (b) the multiplier has to be 2 (16 × 2 = 32)
In (c) the multiplier has to be 5   (8 × 5 = 40)

Two: (a) $\dfrac{4 \times 3 = 12}{5 \times 3 = 15}$ (b) $\dfrac{7 \times 2 = 14}{16 \times 2 = 32}$ (c) $\dfrac{5 \times 5 = 25}{8 \times 5 = 40}$

Raising a fraction to higher terms is necessary in addition and subtraction of fractions. If, for example, you want to add quarters and halves and thirds, you would change them first to equivalent fractions, all having a denominator of 12.

Then, instead of adding $\dfrac{3}{4} + \dfrac{1}{2} + \dfrac{2}{3}$, you would add $\dfrac{9}{12} + \dfrac{6}{12} + \dfrac{8}{12}$.

Not this:

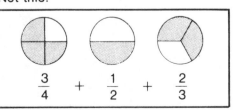

But this:

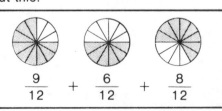

ITEM F–2(II) EQUIVALENT FRACTIONS: REDUCING TO LOWER (AND LOWEST) TERMS

STEPS IN REDUCING TO LOWER (AND LOWEST) TERMS

One: Select a number that divides evenly into both parts.
Two: Divide numerator and denominator by that number.

Example A

Reduce $\frac{15}{20}$ to lower terms.

STEPS

One: The number that divides evenly into both 15 and 20 is 5.

Two: $\dfrac{15 \div 5 = 3}{20 \div 5 = 4}$

$\dfrac{15}{20}$ reduces to $\dfrac{3}{4}$

Example B

Reduce $\frac{26}{65}$ to lower terms.

STEPS

One: 26 and 65 are both divisible by 13.

Two: $\dfrac{26 \div 13 = 2}{65 \div 13 = 5}$

$\dfrac{26}{65} = \dfrac{2}{5}$

You *must distinguish between reducing to* **lower** *terms and reducing to* **lowest** *terms.* Many students lose marks on a test because they do the former when instructions call for the latter. In examples A and B, the fractions have been reduced to lowest terms as well as to lower terms. But consider the next example:

Example C

Reduce $\frac{96}{144}$ to lowest terms.

STEPS

One: There are many common factors of 96 and 144. Both are divisible by 2, 3, 4, 6, 8, 12, 16, 24, or 48. Division by any of these will reduce the fraction to lower terms, but only division by the largest common factor (48) will reduce it to lowest terms.

Two: $\dfrac{96}{144} = \dfrac{96 \div 48}{144 \div 48} = \dfrac{2}{3}$ $\dfrac{96}{144}$ reduces to $\dfrac{2}{3}$

Reducing to lowest terms can be accomplished with a single division, provided you use the *highest* common factor. If you use a common factor which is not the largest possible, you will have to divide two or more times. All the examples below have the same end result but it is reached in different ways.

One division*	Two divisions	More than two divisions
$\dfrac{96 \div 48}{144 \div 48} = \dfrac{2}{3}$	$\dfrac{96 \div 12}{144 \div 12} = \dfrac{8}{12}$	$\dfrac{96 \div 4}{144 \div 4} = \dfrac{24}{36}$
	$\dfrac{8 \div 4}{12 \div 4} = \dfrac{2}{3}$	$\dfrac{24 \div 2}{36 \div 2} = \dfrac{12}{18}$
		$\dfrac{12 \div 3}{18 \div 3} = \dfrac{4}{6}$
		$\dfrac{4 \div 2}{6 \div 2} = \dfrac{2}{3}$

*The best method — if you can do it.

EXERCISE F – 2

1. Raise to higher terms as indicated.
   (a) $\dfrac{5}{12} = \dfrac{?}{96}$ (b) $\dfrac{8}{9} = \dfrac{48}{?}$ (c) $\dfrac{16}{50} = \dfrac{?}{100}$
   (d) $\dfrac{3}{4} = \dfrac{?}{24}$ (e) $\dfrac{9}{5} = \dfrac{36}{?}$ (f) $\dfrac{7}{2} = \dfrac{?}{14}$

2. What is the largest number that will divide evenly into each set?
   (a) 20, 40, and 60   (b) 56 and 96   (c) 120 and 75
   (d) 64, 48, and 32   (e) 3, 4, and 5   (f) 42 and 63

3. In each set of fractions below, two are equivalent and one is different. Which one is different?
   (a) $\dfrac{3}{4}, \dfrac{12}{16}, \dfrac{15}{24}$ (b) $\dfrac{10}{16}, \dfrac{30}{42}, \dfrac{40}{64}$ (c) $\dfrac{25}{100}, \dfrac{50}{150}, \dfrac{75}{300}$ (d) $\dfrac{4}{12}, \dfrac{1}{4}, \dfrac{20}{60}$

4. Which fractions in this set are already in lowest terms?

   $\dfrac{10}{15}, \dfrac{5}{6}, \dfrac{21}{28}, \dfrac{21}{25}, \dfrac{27}{30}, \dfrac{27}{81}, \dfrac{11}{16}, \dfrac{21}{24}$

5. Each of the following fractions has been reduced to lower terms but not to lowest terms. Reduce each to lowest terms.
   (a) $\dfrac{16}{24} = \dfrac{8}{12} = \dfrac{?}{?}$ (b) $\dfrac{125}{1000} = \dfrac{25}{200} = \dfrac{?}{?}$ (c) $\dfrac{63}{144} = \dfrac{21}{48} = \dfrac{?}{?}$ (d) $\dfrac{76}{114} = \dfrac{38}{57} = \dfrac{?}{?}$

6. Supply the missing numerators and denominators that will make each set of fractions equivalent.
   (a) $\dfrac{14}{16} = \dfrac{?}{8} = \dfrac{42}{?}$ (b) $\dfrac{2}{3} = \dfrac{10}{?} = \dfrac{24}{?}$ (c) $\dfrac{20}{?} = \dfrac{5}{6} = \dfrac{?}{72}$ (d) $\dfrac{48}{54} = \dfrac{8}{?} = \dfrac{?}{27} = \dfrac{?}{81}$

7. Reduce to lowest terms:
   (a) $\dfrac{625}{1000}$ (b) $\dfrac{25}{40}$ (c) $\dfrac{21}{98}$ (d) $\dfrac{275}{300}$
   (e) $\dfrac{60}{80}$ (f) $\dfrac{48}{144}$ (g) $\dfrac{24}{64}$ (h) $\dfrac{91}{104}$

ANSWERS FOR EXERCISE F – 2

1. (a) $\frac{40}{96}$ (b) $\frac{48}{54}$ (c) $\frac{32}{100}$ (d) $\frac{18}{24}$ (e) $\frac{36}{20}$ (f) $\frac{49}{14}$

2. (a) 20   (b) 8   (c) 15   (d) 16   (e) 1   (f) 21

3. (a) $\frac{15}{24}$ (b) $\frac{30}{42}$ (c) $\frac{50}{150}$ (d) $\frac{1}{4}$

4. $\frac{5}{6}$, $\frac{21}{25}$, and $\frac{11}{16}$

5. (a) $\frac{2}{3}$ (b) $\frac{1}{8}$ (c) $\frac{7}{16}$ (d) $\frac{2}{3}$

6. (a) $\frac{14}{16} = \frac{7}{8} = \frac{42}{48}$ (b) $\frac{2}{3} = \frac{10}{15} = \frac{24}{36}$ (c) $\frac{20}{24} = \frac{5}{6} = \frac{60}{72}$

   (d) $\frac{48}{54} = \frac{8}{9} = \frac{24}{27} = \frac{72}{81}$

7. (a) $\frac{5}{8}$ (b) $\frac{5}{8}$ (c) $\frac{3}{14}$ (d) $\frac{11}{12}$ (e) $\frac{3}{4}$ (f) $\frac{1}{3}$ (g) $\frac{3}{8}$ (h) $\frac{7}{8}$

ITEM F–3   ADDITION WITH FRACTIONS

STEPS

One: Find the common denominator.
Two: Change fractions to equivalents.
Three: Add whole numbers; add numerators.
Four: Simplify answer
   (a) by changing an improper fraction to a mixed number.
   (b) by reducing to lowest terms.

These steps cover every type of question but often you will find that one or more steps may be omitted.

Example A

Add $\frac{3}{4} + \frac{3}{4}$.

STEPS

One: Not necessary, since the denominator 4 is already common to both.

Two: Not necessary.

Three: $\frac{3}{4} + \frac{3}{4} = \frac{6}{4}$

Four: $\frac{6}{4} = 1\frac{2}{4} = 1\frac{1}{2}$

Example B

Add $10\frac{2}{3} + 4$.

STEPS

One: Not necessary, since there is only one fraction.

Two: Not necessary.

Three: $10 + 4 = 14$
   $\frac{2}{3} + 0 = \frac{2}{3}$

Four: $14 + \frac{2}{3} = 14\frac{2}{3}$

Example C

Add $3\dfrac{5}{8} + 4\dfrac{3}{10} + 9\dfrac{2}{5}$.

STEPS	*Thinking Process*
One: Common denominator is 40	(10, 20, 30, 40)
Two: $3\dfrac{5}{8} = 3\dfrac{25}{40}$	$\dfrac{5 \times 5 = 25}{8 \times 5 = 40}$
$4\dfrac{3}{10} = 4\dfrac{12}{40}$	$\dfrac{3 \times 4 = 12}{10 \times 4 = 40}$
$9\dfrac{2}{5} = 9\dfrac{16}{40}$	$\dfrac{2 \times 8 = 16}{5 \times 8 = 40}$
Three: $16\dfrac{53}{40}$	$3 + 4 + 9 = 16$ $\dfrac{25 + 12 + 16}{40} = \dfrac{53}{40}$
Four: $16\dfrac{53}{40} = 17\dfrac{13}{40}$	$\dfrac{53}{40} = 1\dfrac{13}{40}$

NOTE: Most common denominators can be found mentally. If you have difficulty you might try the following procedure.

1. Select the largest denominator in the set.
2. Mentally "go up" that "times table" until you reach a number into which the other denominators will divide evenly.

For example:
In Example C, the denominators are 8, 10, and 5.
The largest denominator is 10.
The 10-times table is 10, 20, 30, 40, 50, etc.
You can't use 10 or 20 or 30 because 8 will not divide evenly into them.
The first number into which all three denominators will divide is 40.
The common denominator is 40.

Some students prefer the horizontal format for addition with fractions. The next example shows both methods. You should use the method that is easier for you.

Example D

Add $5\frac{5}{6} + 2\frac{2}{3} + 1\frac{7}{12}$

Vertical Format

$$5\frac{5}{6} = 5\frac{10}{12}$$

$$2\frac{2}{3} = 2\frac{8}{12}$$

$$1\frac{7}{12} = 1\frac{7}{12}$$

$$8\frac{25}{12} = 10\frac{1}{12}$$

Horizontal Format

$$5\frac{5}{6} + 2\frac{2}{3} + 1\frac{7}{12}$$

$$= 8\frac{10 + 8 + 7}{12}$$

$$= 8\frac{25}{12}$$

$$= 10\frac{1}{12}$$

EXERCISE F – 3

1. This set involves like fractions (fractions that have the same denominator). Steps *One* and *Two*, therefore, will not be necessary.

   (a) $\frac{2}{3} + \frac{1}{3}$ (b) $\frac{7}{10} + \frac{3}{10} + \frac{5}{10}$ (c) $8\frac{5}{8} + 11\frac{3}{8} + 2\frac{7}{8}$

   (d) $\frac{9}{16} + 3\frac{11}{16} + 12 + 4\frac{5}{16}$ (e) $5\frac{9}{20} + \frac{6}{20}$ (f) $16\frac{5}{9} + \frac{4}{9}$

   (g) $\frac{7}{12} + 4 + 3\frac{5}{12}$ (h) $8 + \frac{9}{32} + \frac{31}{32} + 7$

2. Compare your work for this set with the solutions in the answer section.

   (a) $62\frac{1}{4} + 7\frac{3}{5}$ (b) $3\frac{5}{16} + 8\frac{3}{4} + 6\frac{2}{5} + 3\frac{9}{10}$ (c) $3\frac{1}{3} + 2\frac{5}{6} + 4\frac{7}{9}$

3. Miscellaneous types:

   (a) $4 + 6\frac{5}{8} + \frac{7}{12}$ (b) $\frac{3}{8} + \frac{7}{16} + \frac{1}{2} + \frac{2}{3}$ (c) $\frac{3}{5} + \frac{3}{5}$

   (d) $18 + 6\frac{1}{4} + 10$ (e) $3\frac{2}{5} + 6\frac{3}{4} + 7\frac{5}{12} + 1\frac{4}{15}$

   (f) $14 + 3\frac{3}{5} + 11\frac{5}{9}$ (g) $7\frac{1}{4} + 6\frac{3}{8} + 5\frac{5}{6}$

ANSWERS FOR EXERCISE F – 3

1. (a) 1   (b) $1\frac{1}{2}$ (c) $22\frac{7}{8}$ (d) $20\frac{9}{16}$ (e) $5\frac{3}{4}$ (f) 17   (g) 8   (h) $16\frac{1}{4}$

2.

$$62\frac{1}{4} = 62\frac{5}{20}$$
$$7\frac{3}{5} = 7\frac{12}{20}$$
$$\overline{69\frac{17}{20}}$$

$$3\frac{5}{16} = 3\frac{25}{80}$$
$$8\frac{3}{4} = 8\frac{60}{80}$$
$$6\frac{2}{5} = 6\frac{32}{80}$$
$$3\frac{9}{10} = 3\frac{72}{80}$$
$$\overline{20\frac{189}{80}} = 22\frac{29}{80}$$

$$3\frac{1}{3} = 3\frac{6}{18}$$
$$2\frac{5}{6} = 2\frac{15}{18}$$
$$4\frac{7}{9} = 4\frac{14}{18}$$
$$\overline{9\frac{35}{18}} = 10\frac{17}{18}$$

3. (a) $11\frac{5}{24}$ (b) $1\frac{47}{48}$ (c) $1\frac{1}{5}$ (d) $34\frac{1}{4}$ (e) $18\frac{5}{6}$ (f) $29\frac{7}{45}$ (g) $19\frac{11}{24}$

ITEM F–4 SUBTRACTION WITH FRACTIONS

STEPS

One: *Find the common denominator.*
Two: *Change fractions to equivalents.*
Three: *Regroup (or borrow) if necessary.*
Four: *Subtract whole numbers; subtract numerators.*
Five: *Simplify if necessary.*

Example A

Subtract $\dfrac{3}{4}$ from $4\dfrac{11}{12}$.

STEPS

One: Common denominator is 12

Two: $4\dfrac{11}{12} = 4\dfrac{11}{12}$

$\quad\quad \dfrac{3}{4} = \dfrac{9}{12}$

Four: $4\dfrac{2}{12}$

Five: $4\dfrac{1}{6}$

Example B

Subtract $1\dfrac{3}{4}$ from $4\dfrac{4}{5}$.

STEPS

One: Common denominator is 20

Two: $4\dfrac{4}{5} = 4\dfrac{16}{20}$

$\quad\quad 1\dfrac{3}{4} = 1\dfrac{15}{20}$

Four: $3\dfrac{1}{20}$

Five: Not necessary.

Neither example above required regrouping because, in each case, the fraction in the subtrahend was smaller than the fraction in the minuend. This is not always the case. In examples such as the following, you would have to regroup (or borrow) before you could subtract.

(a) $4\frac{3}{5} - 2\frac{4}{5}$ (you can't take $\frac{4}{5}$ from $\frac{3}{5}$);   change to $3\frac{8}{5} - 2\frac{4}{5}$.

(b) $12 - 3\frac{5}{8}$ (you can't take $\frac{5}{8}$ from $\frac{0}{8}$);   change to $11\frac{8}{8} - 3\frac{5}{8}$.

These are the steps in regrouping or borrowing:

	ONE	TWO	THREE	
Original form	Separate 1 from the whole number	Change this 1 to a fraction	Combine with other fraction	New form
$10\frac{5}{8}$	$(9 + 1) + \frac{5}{8}$	$(9 + \frac{8}{8}) + \frac{5}{8}$	$9 + (\frac{8}{8} + \frac{5}{8})$	$9\frac{13}{8}$
$4\frac{7}{12}$	$(3 + 1) + \frac{7}{12}$	$(3 + \frac{12}{12}) + \frac{7}{12}$	$3 + (\frac{12}{12} + \frac{7}{12})$	$3\frac{19}{12}$
8^*	$(7 + 1)$	$(7 + \frac{6}{6})$	$7 + (\frac{6}{6})$	$7\frac{6}{6}$
$6\frac{2}{9}$	$(5 + 1) + \frac{2}{9}$	$(5 + \frac{9}{9}) + \frac{2}{9}$	$5 + (\frac{9}{9} + \frac{2}{9})$	$5\frac{11}{9}$
$7\frac{3}{5}$	$(6 + 1) + \frac{3}{5}$	$(6 + \frac{5}{5}) + \frac{3}{5}$	$6 + (\frac{5+3}{5})$	$6\frac{8}{5}$

*in a question such as $8 - 4\frac{5}{6}$

The following examples of subtraction show regrouping or borrowing.

(a) $10\frac{1}{5} - 3\frac{4}{5}$	(b) $7\frac{2}{5} - 3\frac{3}{4}$	(c) $16 - 5\frac{5}{8}$	(d) $8\frac{1}{10} - 3\frac{5}{6}$
$10\frac{1}{5} = 9\frac{6}{5}$ $3\frac{4}{5} = 3\frac{4}{5}$ $\overline{6\frac{2}{5}}$	$7\frac{2}{5} = 7\frac{8}{20} = 6\frac{28}{20}$ $3\frac{3}{4} = 3\frac{15}{20} = 3\frac{15}{20}$ $\overline{3\frac{13}{20}}$	$16 = 15\frac{8}{8}$ $5\frac{5}{8} = 5\frac{5}{8}$ $\overline{10\frac{3}{8}}$	$8\frac{1}{10} = 8\frac{3}{30} = 7\frac{33}{30}$ $3\frac{5}{6} = 3\frac{25}{30} = 3\frac{25}{30}$ $\overline{4\frac{8}{30} = 4\frac{4}{15}}$

EXERCISE F – 4

1. Regroup each of the following mixed numbers as a preparation for subtraction.
   (a) $16\frac{1}{5}$ (b) $7\frac{2}{9}$ (c) $3\frac{5}{12}$ (d) $14\frac{1}{3}$ (e) $20\frac{11}{16}$ (f) $9\frac{3}{10}$ (g) $2\frac{23}{25}$

2. Subtract the following. Regrouping will not be necessary.
   (a) $3\frac{3}{4} - 2$ (b) $16\frac{5}{8} - 4$ (c) $11\frac{3}{10} - 6$ (d) $9\frac{15}{32} - 6\frac{7}{32}$ (e) $2\frac{5}{8} - \frac{1}{2}$

3. Regrouping will be required in each of the following:
   (a) $3 - 2\frac{3}{4}$ (b) $16 - 4\frac{5}{8}$ (c) $11\frac{2}{9} - 6\frac{4}{9}$ (d) $36\frac{13}{24} - 20\frac{11}{12}$ (e) $7\frac{1}{10} - 3\frac{5}{6}$

4. Calculate the following.

(a) $14\frac{3}{4}$ $-7\frac{2}{3}$	(b) $16\frac{1}{2}$ -4	(c) 16 $-4\frac{1}{2}$	(d) 200 $-36\frac{5}{9}$	(e) $9\frac{5}{12}$ $-6\frac{7}{12}$	(f) $11\frac{1}{5}$ $-9\frac{7}{8}$
(g) $6\frac{2}{3}$ $-1\frac{5}{9}$	(h) 20 $-6\frac{11}{15}$	(i) $\frac{3}{4}$ $-\frac{5}{12}$	(j) $1\frac{1}{2}$ $-\frac{7}{8}$	(k) $9\frac{4}{15}$ $-2\frac{1}{2}$	(l) $12\frac{7}{32}$ $-8\frac{3}{4}$

ANSWERS FOR EXERCISE F – 4

1. (a) $15\frac{6}{5}$ (b) $6\frac{11}{9}$ (c) $2\frac{17}{12}$ (d) $13\frac{4}{3}$ (e) $19\frac{27}{16}$ (f) $8\frac{13}{10}$ (g) $1\frac{48}{25}$

2. (a) $1\frac{3}{4}$ (b) $12\frac{5}{8}$ (c) $5\frac{3}{10}$ (d) $3\frac{8}{32} = 3\frac{1}{4}$ (e) $2\frac{1}{8}$

3. (a) $\frac{1}{4}$ (b) $11\frac{3}{8}$ (c) $4\frac{7}{9}$ (d) $15\frac{15}{24} = 15\frac{5}{8}$ (e) $3\frac{8}{30} = 3\frac{4}{15}$

4. (a) $7\frac{1}{12}$ (b) $12\frac{1}{2}$ (c) $11\frac{1}{2}$ (d) $163\frac{4}{9}$ (e) $2\frac{10}{12} = 2\frac{5}{6}$ (f) $1\frac{13}{40}$
   (g) $5\frac{1}{9}$ (h) $13\frac{4}{15}$ (i) $\frac{4}{12} = \frac{1}{3}$ (j) $\frac{5}{8}$ (k) $6\frac{23}{30}$ (l) $3\frac{15}{32}$

ITEM F–5   MULTIPLICATION WITH FRACTIONS

STEPS

One: *Convert whole or mixed numbers to improper fractions.*
Two*:   *Cancel wherever possible.*
Three: *Multiply numerators; multiply denominators.*
Four: *Simplify if necessary.*

Example A

Multiply $3\frac{3}{4} \times 4 \times \frac{3}{5} \times 1\frac{3}{8}$.

STEPS

One: $3\frac{3}{4}$ becomes $\frac{15}{4}$; 4 changes to $\frac{4}{1}$; $1\frac{3}{8}$ becomes $\frac{11}{8}$

The question now reads $\frac{15}{4} \times \frac{4}{1} \times \frac{3}{5} \times \frac{11}{8}$

Two:*   Cancel the 4s; cancel 15 with 5.

$$\frac{\overset{3}{\cancel{15}}}{\underset{1}{\cancel{4}}} \times \frac{\overset{1}{\cancel{4}}}{1} \times \frac{3}{\underset{1}{\cancel{5}}} \times \frac{11}{8}$$

Three: Numerators $3 \times 1 \times 3 \times 11 = 99$

Denominators $1 \times 1 \times 1 \times 8 = 8$

Four: $\frac{99}{8} = 12\frac{3}{8}$

A note about cancellation — Cancellation is not absolutely essential but it is a useful short cut. The process is the same as reducing to lowest terms, *i.e.* you divide numerator and denominator by the same number, except that the numerator and denominator may belong to different fractions. With or without cancellation, the final result is the same, but most people find that cancellation makes things easier. Compare the two solutions given for the following question. One uses cancellation, the other does not.

$$\text{Multiply } 5\frac{5}{6} \times \frac{7}{8} \times 4\frac{4}{5} \times \frac{1}{7}$$

With Cancellation   Without Cancellation

$$\frac{\overset{7}{\cancel{35}}}{\underset{1}{\cancel{6}}} \times \frac{\overset{1}{\cancel{7}}}{\underset{2}{\cancel{8}}} \times \frac{\overset{\cancel{4}^{1}}{\cancel{24}}}{\underset{1}{\cancel{5}}} \times \frac{1}{\underset{1}{\cancel{7}}} = \frac{7}{2} = 3\frac{1}{2}$$

$$\frac{35}{6} \times \frac{7}{8} \times \frac{24}{5} \times \frac{1}{7} = \frac{5880}{1680} = 3\frac{84}{168} = 3\frac{1}{2}$$

Consider these additional examples of multiplication.

Find $\frac{3}{5}$ of 25.*	Multiply $5\frac{1}{2}$ by $5\frac{1}{2}$.	$\frac{7}{8} \times 9\frac{3}{5} \times \frac{1}{12}$
$\frac{3}{5} \times \frac{\overset{5}{\cancel{25}}}{1} = \frac{15}{1} = 15$	$\frac{11}{2} \times \frac{11}{2} = \frac{121}{4} = 30\frac{1}{4}$	$\frac{7}{8} \times \frac{\overset{6}{\cancel{48}}}{5} \times \frac{1}{\underset{2}{\cancel{12}}} = \frac{7}{10}$

*NOTE: Whenever the word "of" is used between two numbers, it can be replaced by the multiplication sign.

$$\text{``}\frac{3}{5} \text{ of 25''} \text{ is the same as } \text{``}\frac{3}{5} \times 25\text{.''}$$

$$\text{``}\frac{7}{12} \text{ of 500''} \text{ is equal to } \text{``}\frac{7}{12} \times 500\text{.''}$$

EXERCISE F – 5

Multiply the following:

(a) $\frac{3}{4} \times \frac{4}{9}$ (b) $\frac{2}{3} \times 16$ (c) $\frac{11}{16} \times \frac{5}{12} \times \frac{8}{15}$ (d) $14\frac{2}{5} \times 20$ (e) $7\frac{1}{2} \times \frac{2}{3}$

(f) $12 \times 6\frac{7}{8}$ (g) $3\frac{2}{3} \times 2\frac{5}{8}$ (h) $4\frac{5}{8} \times 6\frac{7}{8} \times 3\frac{1}{3}$ (i) $\frac{2}{5} \times \frac{3}{8} \times 3\frac{1}{3}$

(j) $650 \times \frac{3}{5}$ (k) $\frac{7}{9} \times 9 \times 6$ (l) $\frac{3}{8}$ of 2000   (m) $300 \times \frac{2}{3} \times 15\frac{1}{6}$

(n) Find $\frac{5}{16}$ of 64 000   (o) How much is $\frac{3}{10}$ of 210?

ANSWERS FOR EXERCISE F – 5

Every second question is worked out in full.

(a) $\frac{\overset{1}{\cancel{3}}}{\underset{1}{\cancel{4}}} \times \frac{\overset{1}{\cancel{4}}}{\underset{3}{\cancel{9}}} = \frac{1}{3}$ (b) $10\frac{2}{3}$ (c) $\frac{11}{\underset{2}{\cancel{16}}} \times \frac{\overset{1}{\cancel{5}}}{12} \times \frac{\overset{1}{\cancel{8}}}{\underset{3}{\cancel{15}}} = \frac{11}{72}$ (d) 288

(e) $7\frac{1}{2} \times \frac{2}{3} = \frac{\overset{5}{\cancel{15}}}{\underset{1}{\cancel{2}}} \times \frac{\overset{1}{\cancel{2}}}{\underset{1}{\cancel{3}}} = 5$ (f) $82\frac{1}{2}$ (g) $3\frac{2}{3} \times 2\frac{5}{8} = \frac{11}{\underset{1}{\cancel{3}}} \times \frac{\overset{7}{\cancel{21}}}{8} = \frac{77}{8} = 9\frac{5}{8}$

(h) $105\frac{95}{96}$ (i) $\frac{2}{5} \times \frac{3}{8} \times 3\frac{1}{3} = \frac{\overset{1}{\cancel{2}}}{\underset{1}{\cancel{5}}} \times \frac{\overset{1}{\cancel{3}}}{\underset{4}{\cancel{8}}} \times \frac{\overset{\overset{2}{\cancel{10}}}{\cancel{10}}}{\underset{1}{\cancel{3}}} = \frac{1}{2}$ (j) 390

(k) $\frac{7}{9} \times 9 \times 6 = \frac{7}{\underset{1}{\cancel{9}}} \times \frac{\overset{1}{\cancel{9}}}{1} \times \frac{6}{1} = \frac{42}{1} = 42$ (l) 750

(m) $300 \times \frac{2}{3} \times 15\frac{1}{6} = \frac{\overset{100}{\cancel{300}}}{1} \times \frac{\overset{1}{\cancel{2}}}{\underset{1}{\cancel{3}}} \times \frac{91}{\underset{3}{\cancel{6}}} = \frac{9100}{3} = 3033\frac{1}{3}$ (n) 20 000   (o) 63

ITEM F–6 DIVISION WITH FRACTIONS

STEPS

One: Convert whole or mixed numbers to improper fractions.
Two: Invert the divisor. Change the sign to multiply.
Three: Cancel wherever possible.
Four: Multiply numerators; multiply denominators.
Five: Simplify if possible.

You will notice that the steps in division are identical with those in multiplication with one additional step. There is a reason for this:

Division by any number = Multiplication by the reciprocal of that number.

EXAMPLES

20 divided by 4 is the same as 20 multiplied by $\frac{1}{4}$ $(20 \div 4 = 20 \times \frac{1}{4})$

$3\frac{3}{4}$ divided by 3 is equal to $3\frac{3}{4}$ multiplied by $\frac{1}{3}$ $(3\frac{3}{4} \div 3 = 3\frac{3}{4} \times \frac{1}{3})$

$\frac{2}{5}$ divided by $\frac{2}{10}$ is the same as $\frac{2}{5}$ multiplied by $\frac{10}{2}$ $(\frac{2}{5} \div \frac{2}{10} = \frac{2}{5} \times \frac{10}{2})$

Example A

Divide $3\frac{3}{4}$ by 3.

STEPS

One: $3\frac{3}{4}$ becomes $\frac{15}{4}$; 3 becomes $\frac{3}{1}$ improper fractions

Two: $\frac{15}{4} \div \frac{3}{1} = \frac{15}{4} \times \frac{1}{3}$ (Change sign; invert divisor)

Three: $\frac{\overset{5}{\cancel{15}}}{4} \times \frac{1}{\underset{1}{\cancel{3}}}$ (Cancel 3 with 15)

Four: numerator is 5 (5×1)
       denominator is 4 (4×1)

Five: $\frac{5}{4} = 1\frac{1}{4}$ (Divide 4 into 5)

Study the following additional examples.

	Original form	Change to improper fractions	Change sign; invert; cancel	Multiply num. × num. den. × den.	Simplify
(a)	$4\frac{2}{5} \div 1\frac{2}{9}$	$= \frac{22}{5} \div \frac{11}{9}$	$= \frac{\overset{2}{\cancel{22}}}{5} \times \frac{9}{\underset{1}{\cancel{11}}}$	$= \frac{18}{5}$	$= 3\frac{3}{5}$
(b)	$16 \div 3\frac{1}{4}$	$= \frac{16}{1} \div \frac{13}{4}$	$= \frac{16}{1} \times \frac{4}{13}$	$= \frac{64}{13}$	$= 4\frac{12}{13}$
(c)	$24 \div \frac{1}{4}$	$= \frac{24}{1} \div \frac{1}{4}$	$= \frac{24}{1} \times \frac{4}{1}$	$= \frac{96}{1}$	$= 96$
(d)	$\frac{1}{2} \div \frac{2}{3}$	$=$ no change	$= \frac{1}{2} \times \frac{3}{2}$	$= \frac{3}{4}$	not necessary
(e)	$14\frac{1}{4} \div 3$	$= \frac{57}{4} \div \frac{3}{1}$	$= \frac{\overset{19}{\cancel{57}}}{4} \times \frac{1}{\underset{1}{\cancel{3}}}$	$= \frac{19}{4}$	$= 4\frac{3}{4}$

EXERCISE F – 6

Divide as indicated.

(a) $\frac{1}{2} \div 2$ (b) $16\frac{1}{3} \div 2\frac{1}{3}$ (c) $4\frac{1}{4} \div 4\frac{1}{4}$ (d) $14\frac{3}{4} \div 12$ (e) $16 \div \frac{1}{2}$

(f) $50\frac{1}{4} \div 4\frac{3}{16}$ (g) $13\frac{3}{5} \div 9\frac{1}{15}$ (h) $4\frac{8}{9} \div 8$ (i) $\frac{9}{16} \div \frac{11}{12}$ (j) $6\frac{2}{3} \div \frac{4}{5}$

ANSWERS FOR EXERCISE F – 6

(a) $\frac{1}{4}$ (b) 7   (c) 1   (d) $1\frac{11}{48}$ (e) 32   (f) 12   (g) $1\frac{1}{2}$ (h) $\frac{11}{18}$

(i) $\frac{27}{44}$ (j) $8\frac{1}{3}$

ITEM F–7 TO EXPRESS ONE NUMBER AS A FRACTION OF ANOTHER

STEPS

One: Identify the relevant numbers.
Two: Form a fraction of the two numbers.
Three: Simplify if possible.

Example A

Jean lost 21 pounds on a diet with the result that her new weight was 147 pounds. What fraction of her total original weight did she lose?

SOLUTION STEPS

One: The relevant numbers are 21 and 168 (not 147).

Two: The fraction is $\dfrac{21}{168}$.

Three: $\dfrac{21}{168} = \dfrac{1}{8}$.   She lost $\dfrac{1}{8}$ of her total weight.

Example B

Mr. Jones paid $6650 income tax on a salary of $26 600. What fraction of his salary did he pay in income tax?

SOLUTION STEPS

One: The numbers are 6650 and 26 600.

Two: The fraction is $\dfrac{6650}{26\,600}$.

Three: $\dfrac{6650}{26\,600}$ reduces to $\dfrac{1}{4}$.   Mr. Jones paid $\dfrac{1}{4}$ of his salary in tax.

1/6

EXERCISE F – 7

1. After spending $15.00 for a shirt, Bob had $60.00 left. What fraction of his money did he spend?

2. Express the first number in each pair as a fraction of the second.

   (a) 12, 18 *2/3*   (b) 30, 55 *6/11*   (c) 7 dimes, 3 quarters *14/15*

3. A hairdresser uses a solution of 3 parts of peroxide to 5 parts of tint. What fraction of the solution is peroxide? What fraction is tint? *5/8*

4. A soccer team won 8 games and lost 4. What fraction of the total games played did the team win? *2/3*

5. John received $220.00 interest on a $2000.00 bond. What fraction of the bond's value was the interest? *11/100*

6. Ten years ago there were two women in a welding class that had a total of twenty students. Last year there were six women and fourteen men in a class taught by the same teacher. What fraction of the class was composed of women on each occasion? *1/10 & 3/10*

ANSWERS FOR EXERCISE F – 7

1. He spent $\frac{15}{75}$, or $\frac{1}{5}$, of his money.

2. (a) 12 is $\frac{12}{18}$, or $\frac{2}{3}$, of 18   (b) 30 is $\frac{30}{55}$, or $\frac{6}{11}$, of 55.

   (c) 7 dimes is $\frac{70}{75}$, or $\frac{14}{15}$, of 3 quarters.

3. Peroxide is $\frac{3}{8}$ of the solution; tint is $\frac{5}{8}$.

4. The team won 8 games out of 12, or $\frac{2}{3}$ of the total games played.

5. The interest was $\frac{220}{2000}$, or $\frac{11}{100}$, of the bond's value.

6. Ten years ago $\frac{2}{20}$, or $\frac{1}{10}$, of the class was composed of women.

   Last year $\frac{6}{20}$, or $\frac{3}{10}$, of the class members were women.

ITEM F–8   TO FIND THE WHOLE
WHEN A FRACTION (OF IT) IS KNOWN

Here are some examples of this type of problem.

(a) If $\frac{2}{3}$ of a number is 82, what is the number?

(b) You spent $400.00, which was $\frac{2}{5}$ of your pay cheque. What was the total amount of your cheque?

(c) There were 714 union members in a factory. This was $\frac{7}{8}$ of the total number of employees. How many employees were there?

Solution

Divide the given number by the related fraction.

(a) Divide 82 by $\frac{2}{3}$.   The number is 123.

(b) Divide 400 by $\frac{2}{5}$.   The amount of the cheque was $1000.00.

(c) Divide 714 by $\frac{7}{8}$.   There were 816 employees.

Note the similarity between these problems and whole number problems.

PROBLEMS	SOLUTIONS
(1) If 2 pies contain 4800 calories, how many calories are there in 1 pie? There are 2400 calories in 1 pie.	*Divide* the number of calories by the amount of pie. $4800 \div 2 = 2400$
(2) If $3\frac{1}{2}$ pies contain 8400 calories, how many calories are there in 1 pie? There are 2400 calories in 1 pie.	*Divide* the number of calories by the amount of pie. $8400 \div 3\frac{1}{2} = 2400$
(3) If $\frac{3}{4}$ of a pie contains 1800 calories, how many calories are there in 1 pie? There are 2400 calories in 1 pie.	*Divide* the number of calories by the amount of pie. $1800 \div \frac{3}{4} = 2400$

EXERCISE F – 8

1. If $\frac{1}{2}$ of a number is 67, what is the number?

2. If $\frac{3}{4}$ of a pound of chocolate costs $5.70, how much does a whole pound cost?

3. If you lose $\frac{2}{15}$ of your total weight, or 20 lb., how much did you weigh before the loss?

ANSWERS FOR EXERCISE F – 8

1. The number is 134.   2. It costs $7.60.   3. 150 lb.

ITEM F-9 COMMON ERRORS WITH FRACTIONS

1. REDUCING TO *LOWER* BUT NOT *LOWEST* TERMS

   Example: $\dfrac{84}{96}$ reduces to $\dfrac{21}{24}$, but $\dfrac{21}{24}$ will still reduce to $\dfrac{7}{8}$.

2. CHANGING MIXED NUMBERS TO IMPROPER FRACTIONS IN ADDITION AND SUBTRACTION

   This method will work but it's needlessly complicated and lends itself to error.

   EXAMPLE

   Not recommended:

   $$12\tfrac{11}{12} + 16\tfrac{5}{6} = \tfrac{155}{12} + \tfrac{101}{6} = \tfrac{155}{12} + \tfrac{202}{12} = \tfrac{357}{12} = 29\tfrac{9}{12} = 29\tfrac{3}{4}$$

   Preferred method:

   $$12\tfrac{11}{12} + 16\tfrac{5}{6} = 28\tfrac{11+10}{12} = 28\tfrac{21}{12} = 29\tfrac{9}{12} = 29\tfrac{3}{4}$$

3. LEAVING AN ANSWER UNFINISHED

   Students often complete everything correctly as far as the second-last step but neglect to finish.

   Examples of unfinished answers: $28\tfrac{21}{12}$, $\tfrac{9}{1}$, $2\tfrac{11}{4}$, $156\tfrac{35}{40}$, $\tfrac{289}{25}$.

4. SUBTRACTING A SMALLER FRACTION FROM A LARGER ONE REGARDLESS OF WHICH IS MINUEND AND WHICH IS SUBTRAHEND

   This is a very common mistake.

Incorrect	Correct
$16\tfrac{1}{5}$	$16\tfrac{1}{5} = 15\tfrac{6}{5}$
$-3\tfrac{4}{5}$	$-3\tfrac{4}{5} = \ 3\tfrac{4}{5}$
$13\tfrac{3}{5}$	$12\tfrac{2}{5}$

Incorrect	Correct
14	$14 \ = 13\tfrac{9}{9}$
$-2\tfrac{7}{9}$	$-2\tfrac{7}{9} = \ 2\tfrac{7}{9}$
$12\tfrac{7}{9}$	$11\tfrac{2}{9}$

5. FAILURE TO MAKE USE OF CANCELLATION

   For the solution to $3\tfrac{3}{5} \times 4\tfrac{2}{3} \times 1\tfrac{3}{12}$

   this: $\dfrac{\cancel{18}}{\cancel{5}} \times \dfrac{\cancel{14}}{\cancel{3}} \times \dfrac{\cancel{15}}{\cancel{12}} = 21$

   is much easier than this: $\dfrac{18}{5} \times \dfrac{14}{3} \times \dfrac{15}{12} = \dfrac{3780}{180} = 21$

6. INVERTING THE WRONG NUMBER IN DIVISION

It is only in division that a fraction is inverted and it is always the divisor that is inverted. (The divisor is the number that comes after the sign "÷").

If the question says "÷ 24", you multiply by $\frac{1}{24}$.

If the question says "divide by $\frac{1}{6}$", you multiply by $\frac{6}{1}$.

The dividend (the number before the sign "÷") is never inverted.

$$\text{Never invert dividend} \rightarrow \tfrac{15}{16} \div \tfrac{4}{5} \leftarrow \text{Always invert divisor}$$

7. FAILURE TO ESTIMATE ANSWERS IN MULTIPLICATION AND DIVISION

You should have a mental expectation of the approximate size of an answer before you actually work it out. Then, if the final answer is not close to the estimated answer, you can recognize the unreasonable aspect and take steps to correct it.

EXAMPLES

If you multiply $6\frac{3}{4}$ by $5\frac{1}{3}$, you should expect an answer near 35.

Why? Because $6\frac{3}{4}$ is almost 7 and $5\frac{1}{3}$ is close to 5. $7 \times 5 = 35$.

If you divide 24 by $2\frac{2}{3}$, your answer should be close to 8.

Why? Because $2\frac{2}{3}$ is almost 3, and $24 \div 3 = 8$.

Work the above examples and see how they compare with the estimates.

PRACTICE WITH COMMON FRACTIONS

1. Reduce to lowest terms.
   (a) $\frac{45}{60}$ (b) $\frac{56}{112}$ (c) $\frac{87}{150}$ (d) $\frac{52}{91}$

2. Supply the missing numbers.
   (a) $\frac{3}{4} = \frac{15}{}$ (b) $\frac{5}{} = \frac{35}{56}$ (c) $\frac{21}{70} = \frac{}{10}$

3. Change the improper fractions to mixed numbers, and mixed numbers to fractions.
   (a) $1\frac{2}{3}$ (b) $\frac{11}{4}$ (c) $\frac{15}{7}$ (d) $2\frac{11}{16}$ (e) $\frac{99}{8}$ (f) $21\frac{1}{2}$

4. Perform the operations indicated.
   (a) $3\frac{11}{12} + 5\frac{5}{6}$ (b) $\frac{9}{10} - \frac{4}{5}$ (c) $3\frac{2}{3} \times 30$
   (d) $4\frac{1}{2} \times 2 \div \frac{1}{6}$ (e) $(3\frac{1}{4} - 1\frac{7}{8}) \times 12$ (f) $\frac{2\frac{2}{3}}{16} + 9\frac{3}{4}$
   (g) $15 \div \frac{3}{5} \times 10$ (h) $(\frac{2}{3} \times \frac{3}{5}) \div (\frac{1}{6} \times 3)$

   (i) $\dfrac{5\frac{5}{8} - 3\frac{7}{8}}{1\frac{3}{4}}$ (j) $2\frac{4}{9} + \dfrac{1\frac{1}{4} - \frac{2}{3}}{\frac{7}{8}\text{ of }12}$

PRACTICE WITH COMMON FRACTIONS

1. (a) $\frac{3}{4}$ (b) $\frac{1}{2}$ (c) $\frac{29}{50}$ (d) $\frac{4}{7}$ 2. (a) 20   (b) 8   (c) 3
3. (a) $\frac{5}{3}$ (b) $2\frac{3}{4}$ (c) $2\frac{1}{7}$ (d) $\frac{43}{16}$ (e) $12\frac{3}{8}$ (f) $\frac{43}{2}$
4. (a) $9\frac{3}{4}$ (b) $\frac{1}{10}$ (c) 110   (d) 54   (e) $16\frac{1}{2}$ (f) $9\frac{11}{12}$
   (g) 250   (h) $\frac{4}{5}$ (i) 1   (j) $2\frac{1}{2}$

PRACTICE IN THE FOUR OPERATIONS

This page may be used for practice in several different ways.

For Addition $A + B = C$ $B + D = A$
For Subtraction $A - B = D$ $C - A = B$ $C - B = A$ $A - D = B$
For Multiplication $A \times B = E$ $B \times F = A$
For Division $A \div B = F$ $A \div F = B$ $E \div A = B$ $E \div B = A$

	A	B	C	D	E	F
(a)	$8\frac{1}{3}$	$3\frac{3}{5}$	$11\frac{14}{15}$	$4\frac{11}{15}$	30	$2\frac{17}{54}$
(b)	$4\frac{2}{5}$	$3\frac{3}{4}$	$8\frac{3}{20}$	$\frac{13}{20}$	$16\frac{1}{2}$	$1\frac{13}{75}$
(c)	$8\frac{4}{7}$	$2\frac{7}{10}$	$11\frac{19}{70}$	$5\frac{61}{70}$	$23\frac{1}{7}$	$3\frac{11}{63}$
(d)	$16\frac{1}{5}$	$2\frac{7}{9}$	$18\frac{44}{45}$	$13\frac{19}{45}$	45	$5\frac{104}{125}$
(e)	$3\frac{3}{4}$	$1\frac{5}{6}$	$5\frac{7}{12}$	$1\frac{11}{12}$	$6\frac{7}{8}$	$2\frac{1}{22}$
(f)	$10\frac{5}{12}$	$3\frac{3}{4}$	$14\frac{1}{6}$	$6\frac{2}{3}$	$39\frac{1}{16}$	$2\frac{7}{9}$
(g)	$3\frac{3}{4}$	$3\frac{1}{3}$	$7\frac{1}{12}$	$\frac{5}{12}$	$12\frac{1}{2}$	$1\frac{1}{8}$
(h)	$5\frac{5}{6}$	$2\frac{2}{5}$	$8\frac{7}{30}$	$3\frac{13}{30}$	14	$2\frac{31}{72}$
(i)	$7\frac{11}{12}$	$2\frac{3}{8}$	$10\frac{7}{24}$	$5\frac{13}{24}$	$18\frac{77}{96}$	$3\frac{1}{3}$
(j)	16	$5\frac{5}{8}$	$21\frac{5}{8}$	$10\frac{3}{8}$	90	$2\frac{38}{45}$
(k)	$4\frac{3}{4}$	$1\frac{5}{6}$	$6\frac{7}{12}$	$2\frac{11}{12}$	$8\frac{17}{24}$	$2\frac{13}{22}$
(l)	$4\frac{2}{5}$	$2\frac{3}{4}$	$7\frac{3}{20}$	$1\frac{13}{20}$	$12\frac{1}{10}$	$1\frac{3}{5}$
(m)	$3\frac{3}{8}$	$2\frac{1}{4}$	$5\frac{5}{8}$	$1\frac{1}{8}$	$7\frac{19}{32}$	$1\frac{1}{2}$
(n)	$9\frac{3}{4}$	$5\frac{5}{12}$	$15\frac{1}{6}$	$4\frac{1}{3}$	$52\frac{13}{16}$	$1\frac{4}{5}$

Examples using question (a)

$$8\frac{1}{3} = 8\frac{5}{15}$$
$$+ \ 3\frac{3}{5} = 3\frac{9}{15}$$
$$\overline{ 11\frac{14}{15}}$$

$$8\frac{1}{3} = 8\frac{5}{15} = 7\frac{20}{15}$$
$$- \ 3\frac{3}{5} = 3\frac{9}{15} = 3\frac{9}{15}$$
$$\overline{ 4\frac{11}{15}}$$

$$8\frac{1}{3} \times 3\frac{3}{5} = \frac{\overset{5}{\cancel{25}}}{\underset{1}{\cancel{3}}} \times \frac{\overset{6}{\cancel{18}}}{\underset{1}{\cancel{5}}} = \frac{30}{1} = 30$$

$$8\frac{1}{3} \div 3\frac{3}{5} = \frac{25}{3} \div \frac{18}{5} = \frac{25}{3} \times \frac{5}{18} = \frac{125}{54} = 2\frac{17}{54}$$

FRACTION TEST A
(see page 119.)

Note: Answers must be reduced to lowest terms.

(a) Simplify $\frac{18}{8}$.

(b) $\frac{2}{5} + \frac{2}{5}$

(c) $\frac{7}{8} - \frac{3}{8}$

(d) $\frac{2}{3} + \frac{5}{6}$

(e) $5\frac{5}{6} - 2\frac{2}{3}$

(f) $14\frac{1}{3} - 3\frac{3}{4}$

(g) $6\frac{2}{3} + 5\frac{1}{8} + 3\frac{3}{4}$

(h) 8 is what fraction of 12?

(i) $\frac{7}{10} \times \frac{5}{21}$

(j) $3\frac{3}{4} \times 2\frac{2}{3}$

(k) Find $\frac{2}{3}$ of 48.

(l) $\frac{3}{4} \div 4$

(m) $24 \div 2\frac{1}{2}$

(n) $2\frac{5}{8} \div 3\frac{1}{2}$

(o) Divide 63 by $\frac{7}{9}$.

(p) If $\frac{3}{4}$ of a number is 27, what is the number?

FRACTION TEST B
(see page 119.)

Note: Answers must be reduced to lowest terms.

(a) Simplify $\frac{10}{4}$.

(b) $\frac{3}{8} + \frac{4}{8}$

(c) $\frac{7}{10} - \frac{5}{10}$

(d) $\frac{3}{4} + \frac{5}{8}$

(e) $6\frac{3}{10} + 9\frac{2}{5}$

(f) $19\frac{1}{2} - 5\frac{3}{5}$

(g) $5\frac{3}{8} + 6\frac{1}{3} + 2\frac{3}{4}$

(h) 9 is what fraction of 15?

(i) $\frac{3}{8} \times \frac{12}{15}$

(j) $6\frac{2}{3} \times 3\frac{3}{8}$

(k) Find $\frac{3}{4}$ of 24.

(l) $\frac{5}{12} \div 12$

(m) $36 \div 2\frac{1}{2}$

(n) $4\frac{1}{2} \div 2\frac{1}{4}$

(o) Divide 56 by $\frac{7}{8}$.

(p) If $\frac{2}{3}$ of a number is 56, what is the number?

FRACTION TEST C
(see page 119.)

Note: Answers must be reduced to lowest terms.

(a) Simplify $\frac{14}{4}$.

(b) $\frac{3}{12} + \frac{4}{12}$

(c) $\frac{7}{8} - \frac{3}{8}$

(d) $\frac{2}{3} + \frac{5}{9}$

(e) $6\frac{5}{8} + 3\frac{1}{4}$

(f) $14\frac{1}{4} - 7\frac{5}{8}$

(g) $2\frac{1}{5} + 3\frac{1}{4} + 1\frac{1}{2}$

(h) 6 is what fraction of 10?

(i) $\frac{3}{5} \times \frac{15}{18}$

(j) $3\frac{3}{5} \times 4\frac{1}{4}$

(k) Find $\frac{3}{5}$ of 24.

(l) $\frac{7}{8} \div 4$

(m) $15 \div 1\frac{1}{4}$

(n) $3\frac{3}{5} \div 2\frac{2}{5}$

(o) Divide 42 by $\frac{6}{7}$.

(p) If $\frac{3}{8}$ of a number is 24, what is the number?

For additional practice with fractions see pages 209 to 212.

CHAPTER THREE: **DECIMALS**

The word "decimal" comes from the Latin word that means ten. Our number system is called the decimal number system because it is based upon the number 10.

The decimal number system includes both whole numbers and fractions. These are separated by a decimal point. Whole numbers lie to the left of the decimal point, fractions to the right.

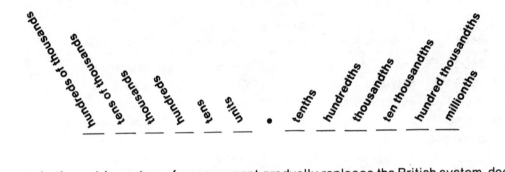

As the metric system of measurement gradually replaces the British system, decimals are becoming increasingly important in school arithmetic, and common fractions less important. This is because the metric system *is* a decimal system; it too is based upon 10 and powers of 10.

ITEMS IN CHAPTER

ITEM D – 1   READING DECIMAL NUMBERS

STEPS

One: Read the whole number (if there is one).
Two: Read the decimal point (it says "and").
Three: Read the numerator as written.
Four: Read the denominator (the place value of the final digit).

EXAMPLES

	Read	Whole number	Decimal point	Numerator	Denominator
(a)	2.020	two	and	twenty	thousandths
(b)	41.06	forty-one	and	six	hundredths
(c)	3000.4	three thousand	and	four	tenths
(d)	6.0058	six	and	fifty-eight	ten-thousandths
(e)	0.000 004	—	—	four	millionths
(f)	1000.019	one thousand	and	nineteen	thousandths
(g)	810.005 25	eight hundred ten	and	five hundred twenty-five	hundred-thousandths
(h)	100.0009	one hundred	and	nine	ten-thousandths*
(i)	0.0109	—	—	one hundred nine	ten-thousandths*

*Note the difference between (h) and (i) although the words appear to be very similar. Only the decimal point should be read as "and".

ANOTHER WAY TO READ DECIMAL NUMBERS

If you were reading a list of numbers, you would probably read this way:

2.020 = two decimal zero two zero
or two point zero two zero

10.658 = ten decimal six five eight
or ten point six five eight

It is a mistake to combine the two methods. For example, to say 1.65 = one point 65 hundredths would be incorrect. It should be either "one and sixty-five hundredths" or "one point six five".

Note: If there is no whole number, a zero is written in the whole number (units) place. This zero is ignored in the reading.

Digits after the decimal point are written in groups of three, just as in whole numbers.

ITEM D-2 WRITING DECIMAL NUMBERS

STEPS

One: Determine the number of places required to indicate the given denominator (see chart below).

Two: Write the given numerator with additional zeros if necessary in order to have the correct number of places.

1 place	denominator is 10	0.4	= 4 tenths
2 places	denominator is 100	0.04	= 4 hundredths
3 places	denominator is 1000	0.004	= 4 thousandths
4 places	denominator is 10 000	0.0004	= 4 ten-thousandths
5 places	denominator is 100 000	0.000 04	= 4 hundred-thousandths
6 places	denominator is 1 000 000	0.000 004	= 4 millionths
		etc.	

Example A

Write six ten-thousandths as a decimal. *0, 000L*

STEPS

One: To indicate ten-thousandths, we need 4 places.

Two: The numerator 6 uses only the last of these places; therefore 3 additional zeros are needed.

Answer is 0.0006 (not 0.6000, which reads "6000 ten-thousandths")

Example B

Write $12\frac{72}{1000}$ in decimal form. *12.092*

STEPS

One: The denominator 1000 requires 3 places.

Two: The numerator 72 uses only two places; one additional zero is needed.

Answer: $12\frac{72}{1000} = 12.072$ (not 12.720)

EXERCISE D – 2

Write each of the following in decimal form:

(a) six tenths   (b) nineteen thousandths   (c) one and five hundredths

(d) three thousand and seventy-two hundredths   (e) eleven millionths

(f) sixty-five ten-thousandths   (g) forty-eight and nine hundredths

(h) nine hundred fifty millionths   (i) nine hundred and fifty millionths

(j) six and three hundred-thousandths   (k) six and three hundred thousandths

ANSWERS FOR EXERCISE D – 2

(a) 0.6   (b) 0.019   (c) 1.05   (d) 3000.72   (e) 0.000 011   (f) 0.0065

(g) 48.09   (h) 0.000 950   (i) 900.000 050   (j) 6.000 03   (k) 6.300

ITEM D-3 CHANGING A DECIMAL FRACTION TO A COMMON FRACTION

STEPS

One: Count the number of digits to the right of the decimal point in order to determine the denominator.

Two: Drop the decimal point and any surplus zeros.

Three: Form a fraction and reduce if possible.

Example A

Change 0.008 to a common fraction in lowest terms.

STEPS

One: The denominator is 1000 (3 places indicate thousandths).

Two: The numerator is 8 (without the zeros).

Three: $\dfrac{8}{1000}$ reduces to $\dfrac{1}{125}$

Example B

Write the fractional equivalent of 16.55

STEPS

One: The denominator is 100.

Two: The numerator is 55.

Three: $\dfrac{55}{100} = \dfrac{11}{20}$ $16.55 = 16\dfrac{11}{20}$ *

Note that the whole number stays the same.

EXERCISE D – 3

1. Write each of the following as a common fraction reduced to lowest terms.

   (a) 0.045   (b) 0.0325   (c) 0.17   (d) 0.002   (e) 0.1875   (f) 0.112

   (g) 0.0367   (h) 0.000 007   (i) 0.8   (j) 0.000 250   (k) 0.000 375

2. Change each of the following to mixed numbers containing common fractions.

   (a) 6.012   (b) 15.4   (c) 200.002   (d) 17.013   (e) 1.625   (f) 25.0075

ANSWERS FOR EXERCISE D – 3

1. (a) $\dfrac{45}{1000} = \dfrac{9}{200}$ (b) $\dfrac{325}{10\,000} = \dfrac{13}{400}$ (c) $\dfrac{17}{100}$ (d) $\dfrac{2}{1000} = \dfrac{1}{500}$ (e) $\dfrac{1875}{10\,000} = \dfrac{3}{16}$

  (f) $\dfrac{112}{1000} = \dfrac{14}{125}$ (g) $\dfrac{367}{10\,000}$ (h) $\dfrac{7}{1\,000\,000}$ (i) $\dfrac{8}{10} = \dfrac{4}{5}$ (j) $\dfrac{1}{4000}$ (k) $\dfrac{3}{8000}$

2. (a) $6\dfrac{3}{250}$ (b) $15\dfrac{2}{5}$ (c) $200\dfrac{1}{500}$ (d) $17\dfrac{13}{1000}$ (e) $1\dfrac{5}{8}$ (f) $25\dfrac{3}{400}$

ITEM D–4   ADDITION AND SUBTRACTION WITH DECIMALS

STEPS

One: *Copy numbers in columns so that whole numbers are lined up with whole numbers, tenths with tenths, etc.*

Two: *Ignore the decimal point momentarily, and add or subtract as you would whole numbers.*

Three: *Insert the decimal point in your answer. Decimal points in question and answer must be in line.*

Examples of addition

A Add 4.3, 2.6, 17, 1.65, and 0.0342.

B $0.6 + 0.3 + 0.1$

C Find the sum of twenty-four, 12.9, four hundredths, and 65.3.

D $8.60 + 77.3 + 4.1$

E What is the sum of 300, 16, and 0.143?

A	B	C	D	E
4.3				
2.6		24		
17	0.6	12.9	8.60	300
1.65	0.3	0.04	77.3	16
0.0342	0.1	65.3	4.1	0.143
25.5842	1.0	102.24	90.00	316.143

Examples of subtraction

A Subtract 0.04 from 17

B $16.2 - 9.32$

C Subtract nine thousandths from 12.7

D Find the difference between 5 and 0.05

A

```
  17        17.00
  0.04  or  0.04
 ─────     ─────
 16.96     16.96
```

B

```
 16.2       16.20
 9.32  or   9.32
 ─────     ─────
 6.88       6.88
```

C

```
 12.7       12.700
 0.009  or  0.009
 ──────     ──────
 12.691     12.691
```

D

```
  5          5.00
 0.05  or   0.05
 ─────      ─────
 4.95        4.95
```

EXERCISE D – 4

1. Copy the questions in vertical form, and add or subtract as indicated.
   (a) 0.3 + 0.2 + 0.5
   (b) 8.7 − 3.05
   (c) seventeen − 14.6
   (d) 200 + 1.6 + 0.007 + 3.5
   (e) Subtract eleven hundredths from seven tenths.
   (f) 0.008 36 + 0.0099
   (g) 17.065 − 3
   (h) 0.164 + 11 − 0.24 + 6
   (i) Subtract 0.08 from 3

2. Add. six and twenty-two hundredths
         twelve and four tenths
         seventeen
         one and seven thousandths
         nineteen hundredths

3. Subtract thirty-four hundredths from two and nine tenths.

4. How much smaller is 0.6 than 10?

5. Express the shaded part of each circle as a decimal fraction of the whole circle.

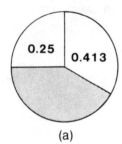

(a)

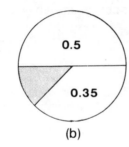

(b)

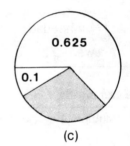

(c)

6. In a three-day period, Tom jogged a total of 40 km. If he ran 15.4 km the first day and 12.7 km the second day, how far did he run on the third day?

7. Correct each of the following:
   (a) 0.4 + 0.5 + 0.3 = 0.12
   (b) 0.62 + 0.14 + 0.38 = 0.114

ANSWERS FOR EXERCISE D – 4

1. (a) 1.0   (b) 5.65   (c) 2.4   (d) 205.107   (e) 0.59
   (f) 0.018 26   (g) 14.065   (h) 16.924   (i) 2.92
2. 36.817   3. 2.56   4. 9.4
5. (a) 0.337   (b) 0.15   (c) 0.275
6. 11.9 km
7. (a) 0.12 should be 1.2   (b) 0.114 should be 1.14

ITEM D–5 MULTIPLICATION WITH DECIMALS

STEPS

One: *Ignore the decimal point momentarily and multiply as you would whole numbers.*

Two: *Count the number of decimal places (digits to the right of the decimal point) in the question.*

Three: *Insert the decimal point in your answer so that the number of places in question and answer are the same. You may have to include additional zeros in order to have the correct number of places.*

Example A

Multiply 0.2 by 0.04.

STEPS

One: $2 \times 4 = 8$

Two: There are three places to the right of the decimal point.

Three: There must be three places in the answer, so two extra zeros are needed in front of the 8.

Answer: $0.2 \times 0.04 = 0.008$

Example B

Multiply $4.5 \times 0.2 \times 0.05$.

STEPS

One: $45 \times 2 \times 5 = 450$

Two: There are four places in the question.

Three: You must have four places in the answer.

Answer: 0.0450 or 0.045

Comment: Most of us multiply decimals in the purely mechanical fashion illustrated above. Perhaps the following will illustrate the reason why the number of places in the question and the answer must match.

Example C

Multiply $0.02 \times 0.4 \times 0.3 \times 2$.

This means $\dfrac{2}{100} \times \dfrac{4}{10} \times \dfrac{3}{10} \times \dfrac{2}{1}$ or $\dfrac{48}{10\,000}$

To show a denominator of 10 000, we need 4 places.

The answer is 0.0048.

Some other examples of multiplication are:

4.22 × 3.6	450 × 0.008	0.215 × 2.4	200 × 0.05
4.22	450	0.215	200
3.6	0.008	2.4	0.05
2532	3.600	860	10.00
1266	or	430	or
15.192	3.6	0.5160	10
		or	
		0.516	

Note: It is not necessary to line up decimal points as it is in addition and subtraction.

EXERCISE D – 5

1. Do not multiply the following; instead write the total number of decimal places that there would be in each answer.

   (a) 0.4 × 0.06 × 24   (b) 350.2 × 1.068

   (c) 1400 × 1300   (d) 457 × 28.1

   (e) 5 × 62 × 3.4 × 0.016   (f) 0.05 × 0.005

2. Multiply.

   (a) 600 × 0.04   (b) 0.4 × 0.04 × 4.2

   (c) 16.5 × 8.1   (d) 0.029 × 1.2

   (e) 3.2 × 120 × 0.003   (f) 0.3 × 0.3 × 0.3

   (g) 124.6 × 0.51   (h) 0.005 × 0.007

3. Multiply each set and find the sum of the products.

   (a)   175 × 0.4   (b)   65 × 2
        17.5 × 0.04   0.065 × 2
        1.75 × 4   65 × 0.002
        0.175 × 0.004   0.65 × 200

ANSWERS FOR EXERCISE D – 5

1. (a) 3 places   (b) 4 places   (c) 0 places

   (d) 1 place   (e) 4 places   (f) 5 places

2. (a) 24   (b) 0.0672   (c) 133.65   (d) 0.0348

   (e) 1.1520 or 1.152   (f) 0.027   (g) 63.546   (h) 0.000 035

3. (a) 70 + 0.7 + 7 + 0.0007 = 77.7007

   (b) 130 + 0.13 + 0.13 + 130 = 260.26

ITEM D – 6 DIVISION WITH DECIMALS

STEPS

One: Make the divisor a whole number (by moving the decimal point the required number of places).

Two: Move the decimal point in the dividend the same number of places.

Three: Place the decimal point in its new location in the quotient.

Four: Proceed to divide, placing each digit in the quotient exactly above the corresponding digit in the dividend.

Note: Since decimal division causes more errors than the other three operations combined, we will examine this topic in a little more detail than we have the previous topics.

EXAMPLES OF STEP ONE

Divisor is 0.04)	Move decimal point 2 places	Divisor is now 4)
Divisor is 26)	Already a whole number	No change
Divisor is 1.765)	Move decimal point 3 places	Divisor is now 1765)
Divisor is 2400.2)	Move decimal point 1 place	Divisor is now 24002)

EXAMPLES OF STEPS ONE, TWO, AND THREE

Original Form	Change	New Form	Step Three*
0.04) 350.2	0.04) 350.20	4) 35020	4) 35020.
0.2) 17.53	0.2) 17.53	2) 175.3	2) 175.3
400) 0.027	No change	No change	400) 0.027
0.0004) 1.6	0.0004) 1.6000	4) 16000	4) 16000.
50) 5	No change	No change	50) 5.

Note: The decimal point in the quotient is exactly above the decimal point in the dividend.

MORE EXAMPLES

Original Form	Revised Form	Thinking Process to Obtain Quotient	Answer
0.4) 0.016	4) 0.16	4 into 0 goes 0 times 4 into 1 goes 0 times 4 into 16 goes 4 times	0.04 4) 0.16
500) 5	No change	500 into 5 goes 0 times 500 into 50 goes 0 times 500 into 500 goes 1 time	0.01 500) 5.00
0.2) 240	2) 2400		1200 2) 2400
0.15) 0.5	15) 50	15 into 50 goes 3 times 15 into 50 goes 3 times 15 into 50 goes 3 times This will never work out evenly.	3.33 * 15) 50.00 45 5 0 4 5 50 45 5 etc.

The dot indicates that this is a recurring or repeating decimal.

SOME QUESTIONS THAT YOU MIGHT ASK:

Q. Why make the divisor a whole number?

A. To simplify the operation. It's easier to deal with a single decimal number than to juggle the value of two mixed numbers. In fact, some division (eg. the European method) eliminates the decimal point in both dividend and divisor.

EXAMPLE

0.04) 0.028 becomes 40) 28 rather than 4) 2.8

0.2) 0.0016 becomes 2000) 16 rather than 2) 0.016

The end result is the same, but the European method is less preferred because of the large, unwieldy divisors involved.

Q. How can you arbitrarily move decimal points and still get the right answer?

A. This is an application of the principle of equivalent fractions.
A fraction is a statement of division.

$\frac{15}{4}$ means $15 \div 4$; conversely, $20 \div 0.02$ is the same as $\frac{20}{0.02}$

If we multiply each part of the fraction $\frac{20}{0.02}$ by 100, it becomes $\frac{2000}{2}$.

This happens in division when 0.02) 20 is changed to 2) 2000

Q. Why is it OK to add any number of zeros after a decimal point?

A. Again, it's the principle of equivalent fractions.

$$0.1 = \frac{1}{10} \qquad 0.10 = \frac{10}{100}, \text{which reduces to } \frac{1}{10}.$$

$$0.1000 = \frac{1000}{10\,000}, \text{which also reduces to } \frac{1}{10}.$$

Similarly, $0.42 = 0.420 = 0.42000 = 0.4200000\ldots$

Any of these will reduce to $\dfrac{42}{100}$.

Q. How can I prove an answer in decimal division?

A. The same as in whole number division. If you multiply the quotient times the divisor, the result, if correct, will be equal to the dividend.

Division	Proof by Multiplication		
$0.04\overline{)1.6} = 4\overline{)160}^{\,40}$	Quotient ×Divisor Dividend	40 ×0.04 1.60	or 1.6
$20\overline{)0.05}^{\,0.0025}$	Quotient ×Divisor Dividend	0.0025 ×20 0.0500	or 0.05

EXERCISE D – 6

Calculate and prove each answer. Then compare your work with the solutions below.

1. (a) $2400 \div 0.06$ (b) $2.4 \div 600$ (c) $0.024 \div 60$
   (d) $240 \div 0.006$ (e) $24 \div 60$

	Original Form	Revised Form	Proof by Multiplication
(a)	$2400 \div 0.06$	$6\overline{)240\,000}$ quotient $40\,000$	$40\,000 \times 0.06 = 2400$
(b)	$2.4 \div 600$	$600\overline{)2.400}$ quotient 0.004	$0.004 \times 600 = 2.4$
(c)	$0.024 \div 60$	$60\overline{)0.0240}$ quotient 0.0004	$0.0004 \times 60 = 0.024$
(d)	$240 \div 0.006$	$6\overline{)240\,000}$ quotient $40\,000$	$40\,000 \times 0.006 = 240$
(e)	$24 \div 60$	$60\overline{)24.0}$ quotient 0.4	$0.4 \times 60 = 24$

2. (a) $375 \div 25$ (b) $37.5 \div 250$ (c) $0.375 \div 0.25$
   (d) $37.5 \div 0.025$ (e) $0.375 \div 25$

3. (a) $60 \div 0.015$ (b) $0.6 \div 150$ (c) $6 \div 0.015$
   (d) $6000 \div 15$ (e) $0.0006 \div 1.5$

4. (a) $24 \div 30$ (b) $0.024 \div 0.3$ (c) $2400 \div 0.03$
   (d) $2.4 \div 3000$ (e) $0.0024 \div 300$

Note: There is a difference between "divide by" and "divide into."
   "Divide 5 by 10" = $10\overline{)5}$ *but* "Divide 5 into 10" = $5\overline{)10}$
   "3 divided into 72" = $3\overline{)72}$ *but* "3 divided by 72" = $72\overline{)3}$

5. Find the quotient when 40 is divided into 12.
6. Divide 0.08 by 16.
7. Find the quotient when 2.4 is divided into 0.096.
8. Find each quotient.

   (a) $\dfrac{402}{0.02}$ (b) $\dfrac{600}{0.06}$ (c) $\dfrac{0.35}{70}$ (d) $\dfrac{550}{1.1}$

ANSWERS FOR EXERCISE D – 6

2. (a) 15   (b) 0.15   (c) 1.5   (d) 1500   (e) 0.015
3. (a) 4000   (b) 0.004   (c) 400   (d) 400   (e) 0.0004
4. (a) 0.8   (b) 0.08   (c) 80 000   (d) 0.0008   (e) 0.000 008
5. 0.3   6. 0.005   7. 0.04
8. (a) 20 100   (b) 10 000   (c) 0.005   (d) 500

ITEM D – 7 SHORT-CUT MULTIPLICATION
AND DIVISION BY A POWER OF 10

To multiply or divide by a power of 10, you move the decimal point the number of places indicated by the exponent, or the number of places that there are zeros in the multiplier or divisor.

The following example shows that these are the same.

Power	Value of power	Exponent	Number of zeros in the power	Move the decimal point
10^1	10	1	1	1 place
10^2	100	2	2	2 places
10^3	1000	3	3	3 places
10^4	10 000	4	4	4 places
10^5	100 000	5	5	5 places
10^6	1 000 000	6	6	6 places

Multiplication by a power of 10 results in a larger answer.
To multiply by a power of 10, you move the decimal point to the *right*.

Division by a power of 10 results in a smaller answer.
To divide by a power of 10, you move the decimal point to the *left*.

EXAMPLES

Problem	Number of places	Direction to move	Answer
14.2×10	1 zero; 1 place	right (multiply)	142
$0.16 \div 100$	2 zeros; 2 places	left (divide)	0.0016
$1242 \div 10$	1 zero; 1 place	left (divide)	124.2
0.136×1000	3 zeros; 3 places	right (multiply)	136
0.0142×10^6	exponent 6; 6 places	right (multiply)	14 200
$15 \div 10^4$	exponent 4; 4 places	left (divide)	0.0015

Note: Do not confuse moving the decimal point with adding zeros. In multiplying 0.41×1000, for example, you move the decimal point three places; however, you add only one zero because the 4 and the 1 account for two of the three places.

$$0.41 \times 1000 = 0.410 = 410$$

$$34.8 \div 1000 = 034.8 = 0.0348$$

EXAMPLES OF MULTIPLICATION BY A POWER OF 10

$0.014 \times 100 = 0.014 = 1.4$	(2 zeros in 100; move 2 places)
$16.5 \times 1000 = 16.500 = 16\ 500$	(3 zeros in 1000; move 3 places)
$1.7 \times 10^4 = 1.7000 = 17\ 000$	(exponent is 4; move 4 places)
$0.0016 \times 10 = 0.0016 = 0.016$	(1 zero in 10; move 1 place)

EXAMPLES OF DIVISION BY A POWER OF 10

$0.014 \div 100 = 00.014 = 0.000\ 14$	(2 zeros in 100; move 2 places)
$16.5 \div 1000 = 016.5 = 0.0165$	(3 zeros in 1000; move 3 places)
$1.7 \div 10^4 = 0001.7 = 0.000\ 17$	(exponent is 4; move 4 places)
$0.0016 \div 10 = 0.0016 = 0.000\ 16$	(1 zero in 10; move 1 place)

EXERCISE D – 7

Cover the answers while you write your own answers for each of the following, then check.

Multiplication		Answers
23.5×100	=	2350
2×10^4	=	20 000
0.254×1000	=	254
14×10^6	=	14 000 000
0.025×10	=	0.25
100.05×10^3	=	100 050

Division		Answers
$23.5 \div 100$	=	0.235
$2 \div 10^4$	=	0.0002
$0.254 \div 1000$	=	0.000 254
$14 \div 10^6$	=	0.000 014
$0.025 \div 10$	=	0.0025
$100.05 \div 10^3$	=	0.100 05

MORE PRACTICE IN MULTIPLICATION AND DIVISION

(a) 7.4×1000

(b) $0.628 \div 10$

(c) $387 \div 10^4$

(d) 0.3598×100

(e) 2.86×10^3

(f) $10\ 000 \div 10^5$

(g) $6 \div 10\ 000$

(h) $65\ 273 \times 10^2$

ANSWERS

(a) 7400 (b) 0.0628 (c) 0.0387 (d) 35.98

(e) 2860 (f) 0.1 (g) 0.0006 (h) 6 527 300

ITEM D – 8   ROUNDING (OFF) DECIMALS

Rounding (or rounding off) is used in decimals as well as in whole numbers to provide an approximate rather than an exact answer. Consider the following examples.

$10.65 lies between $10 and $11. It is closer to $11 than it is to $10. Therefore, $10.65 rounded to the nearest dollar is $11.

224 lies between 200 and 300. It is closer to 200. Rounded to the nearest hundred, 224 becomes 200.

65.317 is between 65.310 and 65.320. It is closer to 65.320. Rounded to two places, 65.317 becomes 65.32. The same number (65.317) is greater than 65.3 but less than 65.4. Rounded to one place, 65.317 becomes 65.3.

STEPS IN ROUNDING NUMBERS

One: *Determine the number of places in the answer.*

Two: *Examine the digit in the "next" place. (If you want 2 places in the answer, examine the third place; if you want 4 places in the answer, examine the fifth place, etc.)*

Three: If this digit is 5 or greater, the final digit in the answer rounds upward; if it's less than 5, the final digit doesn't change.

Example A

Round 42.937 to two places.

STEPS

One: The answer will be either 42.93 (as written) or 42.94.

Two: The third-place digit is 7, which is greater than 5.

Three: Answer is 42.94 (the final digit rounded upward because of the 7)

Example B

Round 0.58312 to the nearest tenth.

STEPS

One: The answer will be either 0.5 (written) or 0.6

Two: The second-place digit is 8, which is greater than 5.

Three: Answer is 0.6 (rounded upwards because of the digit 8).

Example C

Change $\dfrac{1}{12}$ to a three-place decimal.

STEPS

One: The answer will be either 0.083 or 0.084.

Two: The fourth-place digit is 3, which is less than 5.

Three: Answer is 0.083 (doesn't change because 3 is less than 5).

$$\overset{0.08333}{12\overline{)\,1.00000}} \text{ etc.}$$

MORE EXAMPLES

The guide words at the left may help you to think through the steps. Be sure to read down.

	A	B	C
The Number	17.942	0.3359	15.047
Round to	2 places	3 places	nearest tenth
Either or	17.94 or 17.95	0.335 or 0.336	15.0 or 15.1
Deciding digit	2 (less than 5)	9 (more than 5)	4 (less than 5)
Change or not	no change	5 changes to 6	no change
Answer	17.94	0.336	15.0

EXERCISE D – 8

1. Here is a similar set partially completed. Fill in blanks and give answers on bottom line.

	A	B	C
The Number	0.53921	1600.347	2.999
Round to	nearest thousandth	2 places	1 place
Either or		1600.34 or 1600.35	
Deciding digit	2		9
Change or not		change	
Answer			

More practice in rounding:

2. Round to the nearest tenth.
   (a) 5.17   (b) 0.064   (c) 0.243   (d) 20.995
3. Round to two places.
   (a) 5.614   (b) 0.0391   (c) 0.333   (d) 5.555
4. Round to the nearest ten-thousandth.
   (a) 0.647398   (b) 25.9372210
5. Round to the nearest whole number.
   (a) 1.6   (b) 253.29   (c) 17.158

ANSWERS FOR EXERCISE D – 8

1. (a) 0.539   (b) 1600.35   (c) 3.0
2. (a) 5.2   (b) 0.1   (c) 0.2   (d) 21.0
3. (a) 5.61   (b) 0.04   (c) 0.33   (d) 5.56
4. (a) 0.6474   (b) 25.9372
5. (a) 2   (b) 253   (c) 17

ITEM D – 9 CHANGING A COMMON FRACTION TO A DECIMAL

Since a common fraction implies division ($\frac{5}{8}$ means $5 \div 8$), it can be converted to its decimal equivalent by dividing denominator into numerator.

STEPS

One: *Copy the fraction in the form of division.*

Two: *Place decimal points in dividend and quotient.*

Three: *Divide until answer works out evenly, or to three places unless instructed otherwise.*

EXAMPLES: Change to decimal form.   (a) $\dfrac{1}{40}$ (b) $\dfrac{5}{9}$ (c) $5\dfrac{3}{16}$

(a)	(b)	(c)
$\begin{array}{r} 0.025 \\ 40)\overline{1.000} \\ \underline{80} \\ 200 \\ \underline{200} \\ 0 \end{array}$	$\begin{array}{r} 0.55\dot{5} \\ 9)\overline{5.000} \\ \underline{45} \\ 50 \\ \underline{45} \\ 50 \\ \underline{45} \\ 5 \end{array}$	$\begin{array}{r} 0.1875 \\ 16)\overline{3.0000} \\ \underline{1\,6} \\ 1\,40 \\ \underline{1\,28} \\ 120 \\ \underline{112} \\ 80 \\ \underline{80} \\ 0 \end{array}$
$\dfrac{1}{40} = 0.025$	$\dfrac{5}{9} = 0.55\dot{5}$ or 0.556	$5\dfrac{3}{16} = 5.1875$

Note: In example (a), do not omit the zero.

In example (b), the answer is a recurring decimal; it will never work out evenly.

In example (c), the whole number does not change, nor does it affect the calculation.

EXERCISE D – 9

1. Change each fraction in the following set to its decimal equivalent.

   $\dfrac{1}{4}$ $\dfrac{2}{5}$ $\dfrac{9}{10}$ $\dfrac{5}{16}$ $\dfrac{2}{8}$ $4\dfrac{7}{8}$ $9\dfrac{1}{32}$ $2\dfrac{11}{40}$ $32\dfrac{7}{25}$

2. Change each of the following to a decimal correct to 3 places.

   $\dfrac{2}{3}$ $\dfrac{5}{6}$ $\dfrac{9}{11}$ $\dfrac{1}{15}$ $\dfrac{7}{30}$ $\dfrac{5}{12}$

ANSWERS FOR EXERCISE D – 9

1. 0.25   0.4   0.9   0.3125   0.25   4.875   9.031 25   2.275   32.28

2. 0.667   0.833   0.818   0.067   0.233   0.417

ITEM D – 10   MIXED OPERATIONS
WITH FRACTIONS AND DECIMALS

Sometimes a problem contains both common and decimal fractions. It may or may not specify which form to use. Some people prefer to use decimals for all calculations; others are more at ease with common fractions. It is good to be versatile enough that you can select the best form to use in each particular instance.

Having decided which form to use, you must convert as necessary so that all numbers are in the same form.

EXAMPLES

Multiply $\dfrac{5}{12}$ by 7.5.		
Decimal Solution		*Fraction Solution*
$\dfrac{5}{12} = 0.41\dot{6}$ or 0.417 $\begin{array}{r} 0.417 \\ 7.5 \\ \hline 2085 \\ 2919 \\ \hline 3.1275 \end{array}$ or 3.128	In this example with a non-terminating decimal, it is easier to get an exact answer using common fractions. The decimal answer is approximate.	$7.5 = 7\frac{1}{2}$ $\frac{5}{12} \times 7\frac{1}{2}$ $\frac{5}{\cancel{12}} \times \frac{\cancel{15}^{5}}{2} = \frac{25}{8} = 3\frac{1}{8}$
Add $\dfrac{3}{8} + 0.4 + 16\dfrac{1}{2}$		
Decimal Solution		*Fraction Solution*
$\begin{array}{r} \frac{3}{8} = \quad 0.375 \\ 0.4 \\ 16\frac{1}{2} = 16.5 \\ \hline 17.275 \end{array}$	For most people the decimal solution is easier in this case.	$\frac{3}{8} = \frac{15}{40}$ $0.4 = \frac{4}{10} = \frac{16}{40}$ $16\frac{1}{2} = 16\frac{20}{40}$ $16\frac{51}{40} = 17\frac{11}{40}$

EXERCISE D – 10

1. Solve each of the following using both methods. Decide which method seems to be easier in each case.

   (a) $0.4 \times \frac{4}{100}$ (b) $0.6 - \frac{3}{5}$ (c) Subtract 0.5 from $\frac{7}{8}$

   (d) $4\frac{3}{10} + 2.5$ (e) Find $\frac{3}{8}$ of 0.62   (f) Divide $\frac{3}{4}$ by 0.075

2. Work these out using decimals.

   (a) $200 + 0.314 + \frac{4}{25} + 3\frac{1}{10}$ (b) $22.6 \times \frac{3}{4}$ (c) Divide $3\frac{1}{4}$ by 0.25

ANSWERS FOR EXERCISE D – 10

1. (a) $\frac{2}{125}$ or 0.016   (b) 0   (c) $\frac{3}{8}$ or 0.375   (d) $6\frac{4}{5}$ or 6.8

   (e) $\frac{93}{400}$ or 0.2325   (f) 10

2. (a) 203.574   (b) 16.95   (c) 13

ITEM D–11 COMMON ERRORS IN DECIMALS

1. FAILURE TO RECOGNIZE A WHOLE NUMBER

   If a number has no decimal point, it is a whole number.

	Incorrect	Correct
This error occurs in addition. Add 20.3 + 16 + 1.024	20.3 .16 +1.024	20.3 16 +1.024
It occurs in subtraction. Subtract 0.12 from 15.	15 −.12	15.00 −0.12
It occurs most of all in division. Divide 16 by 0.04.	.04 .04) 16	400 0.04) 16.00

2. FAILURE TO CARRY IN ADDITION

   (a) $0.2 + 0.5 + 0.4$ (b) $0.52 + 0.36 + 0.21$

These are incorrect.		These are correct.	
.2	.52	0.2	0.52
.5	.36	0.5	0.36
.4	.21	0.4	0.21
.11	.109	1.1	1.09

3. FAILURE TO DIFFERENTIATE BETWEEN ESSENTIAL AND NON-ESSENTIAL ZEROS

Zeros in decimal numbers are superfluous and may be crossed out in two situations:

(a) When the zero is at the extreme left of a whole number.

0650 is the same as 650.
It is OK to do this: Ø650

(b) When the zero is at the extreme right of a decimal.

0.4200 is equal to 0.42.
It is OK to do this: 0.42ØØ

Zeros have a purpose and cannot be eliminated in two cases:

(a) When zero is the final digit in a whole number.

650 is not equal to 65.
It is *not* OK to do this: 65Ø

(b) When the zero(s) is between the decimal point and the other digits on the decimal side.

0.0042 is not equal to 0.42.
You must *not* do this: 0.ØØ42

Summary:

4. OMISSION OF NECESSARY ZEROS IN A QUOTIENT

This is one of the major errors in decimal division. It can be avoided by placing the decimal point in the quotient directly above the decimal point in the dividend, and then making sure that each digit after the decimal point has a corresponding digit in the quotient.

$$0.16 \text{ divided by } 40 = 40)\overline{0.16} = \begin{array}{r} 0.004 \\ 40)\overline{0.160} \end{array}$$

5. MOVING THE DECIMAL POINT THE WRONG DIRECTION IN SHORT CUT MULTIPLICATION AND DIVISION BY A POWER OF 10

Is 4.62 multiplied by 100 equal to 0.0462 or 462?

Remember that multiplication by a whole number gives a larger result. Division by a whole number produces a smaller result.

$$6.15 \times 1000 = 6150$$
$$6.15 \div 1000 = 0.00615$$

PRACTICE WITH DECIMALS

1. This exercise with its self-contained answers will provide practice in the four operations.

For Addition	$A + B = C$	$B + D = A$	
For Subtraction	$A - B = D$	$C - B = A$	$C - A = B$ $A - D = B$
For Multiplication	$A \times B = E$	$B \times F = A$	
For Division	$A \div B = F$	$E \div A = B$	$E \div B = A$ $A \div F = B$

	A	B	C	D	E	F
(a)	18	0.09	18.09	17.91	1.62	200
(b)	4.3	2	6.3	2.3	8.6	2.15
(c)	20	1.6	21.6	18.4	32	12.5
(d)	14	0.7	14.7	13.3	9.8	20
(e)	2.2	1.1	3.3	1.1	2.42	2
(f)	30	1.5	31.5	28.5	45	20
(g)	24.8	16	40.8	8.8	396.8	1.55
(h)	9	0.045	9.045	8.955	0.405	200
(i)	2.5	0.06	2.56	2.44	0.15	41.667

Example using question (a)

$18 + 0.09 = 18.09$ $18 \times 0.09 = 1.62$

$18.00 - 0.09 = 17.91$ $18 \div 0.09 = 200$

2. In each of the following, change the common fraction to a decimal fraction and solve:

 (a) $5\frac{3}{4} + 2.7 + 9.03$ (b) $11\frac{1}{2} - 2.465$

 (c) Find $\frac{3}{10}$ of 1.905   (d) $0.25 \div \frac{1}{10}$

3. Round the decimal equivalent for each of the following to 3 places.

 $\frac{7}{9}$ $\frac{2}{3}$ $\frac{1}{7}$ $\frac{11}{32}$ $\frac{19}{24}$

ANSWERS FOR PRACTICE QUESTIONS 2 AND 3

2. (a)   5.75   (b) 11.500   (c) 1.905   (d)

       2.7   2.465   0.3

       9.03   9.035   0.5715

      17.48

$$0.1\overline{)0.25} \quad \frac{2.5}{}$$

3. 0.778   0.667   0.143   0.344   0.792

PRACTICE WITH DECIMALS

1. Write as numerals.

 (a) two hundred and twenty thousandths   (b) sixteen millionths

 (c) seventy-five and one hundred forty ten-thousandths

2. Add or subtract as indicated.

    (a) $0.2 + 0.8$ (b) 7 tenths $-$ 7 thousandths   (c) $200 - 25.907$

    (d) Subtract 0.004 from one and 12 thousandths   (e) $2.5 - 0.035$

    (f) Which is larger and by how much: 0.05 or 0.0497?

    (g) $200 + 167.95 + 0.985 + 16 + 543.275$

3. Change to common fractions in lowest terms.

    (a) 0.675   (b) $0.08\dot{3}$ (c) 0.075   (d) 0.84   (e) 0.9875   (f) 0.400

4. (a) Round to the nearest tenth. (i) 17.646   (ii) 0.328   (iii) 100.875

    (b) Round to 3 places. (i) $0.66\dot{6}$ (ii) $0.08\dot{3}$ (iii) 0.1416

    (c) Round to the nearest whole number. (i) 17.75   (ii) 1.05   (iii) 6.6

5. Change to equivalent decimal fractions.

    (a) $\frac{4}{5}$ (b) $\frac{1}{100}$ (c) $16\frac{27}{40}$ (d) $\frac{7}{500}$ (e) $4\frac{11}{15}$ (f) $\frac{280}{1000}$ (g) $2\frac{9}{16}$

6. Multiply:

    (a) 0.043×0.2 (b) $100 \times 2.4 \times 0.02$ (c) 175×0.010

    (d) $0.2 \times 0.3 \times 0.005$ (e) 2000×2.245 (f) 0.02^3

7. Divide; then add the quotients.

    (a) $148 \div 0.04$ (b) $1.44 \div 0.12$ (c) $65 \div 13$
    $\ 1.48 \div 400$ $14.4 \div 1200$ $6500 \div 0.13$
    $\ 1.48 \div 0.4$ $0.144 \div 0.12$ $6.5 \div 130$
    $\ 148 \div 400$ $1.44 \div 1.2$ $0.0065 \div 1.3$
    $\ +0.00148 \div 4$ $+0.0144 \div 0.012$ $+6.5 \div 1300$

8. Remember to use the correct order of operations.

    (a) $0.3 + 0.5 \times 0.6$ (b) $(1.2 + 3.4) \div (2 \times 0.023)$

ANSWERS

1. (a) 200.020   (b) 0.000 016   (c) 75.0140

2. (a) 1.0   (b) 0.693   (c) 174.093   (d) 1.008   (e) 2.465

    (f) 0.05 is larger by 0.0003   (g) 928.21

3. (a) $\frac{27}{40}$ (b) $\frac{1}{12}$ (c) $\frac{3}{40}$ (d) $\frac{21}{25}$ (e) $\frac{79}{80}$ (f) $\frac{2}{5}$

4. (a) (i) 17.6   (ii) 0.3   (iii) 100.9

    (b) (i) 0.667   (ii) 0.083   (iii) 0.142

    (c) (i) 18   (ii) 1   (iii) 7

5. (a) 0.8   (b) 0.01   (c) 16.675   (d) 0.014   (e) 4.733   (f) 0.280   (g) 2.5625

6. (a) 0.0086   (b) 4.8   (c) 1.75   (d) 0.0003   (e) 4490   (f) 0.000 008

7. (a) 3700   (b) 12   (c) 5   8. (a) 0.6   (b) 100
    $\ 0.003\ 7$ 0.012   50 000
    $\ 3.7$ 1.2   0.05
    $\ 0.37$ 1.2   0.005
    $\ +0.000\ 37$ $+1.2$ $+0.005$

    $\ 3704.074\ 07$ 15.612   50 005.060

DECIMAL TEST A
(see page 119.)

1. Write decimal equivalents for the following.
   (a) seven tenths   (b) thirty-three thousandths   (c) $\frac{15}{16}$
2. Which is the largest: 3.6, 3.66, 0.3366, or 3.036?
3. Round 0.4092 to the nearest hundredth.
4. Change 0.35 to a common fraction in lowest terms.
5. $0.2 + 0.5 + 0.3$ 6. $0.016 - 0.008$ 7. $3.4 + 20 + 0.07$
8. $\frac{3}{4} + 0.06 + 2.9$ 9. $22.07 - 1.306$ 10. Take 0.15 from 34.2
11. Multiply 0.076 by 0.032   12. Multiply 0.3 by 0.06 by 20
13. $6)\overline{0.003}$ 14. Divide 32 by 0.4.

DECIMAL TEST B
(see page 119.)

1. Write decimal equivalents for the following.
   (a) three tenths   (b) sixty-one thousandths   (c) $\frac{9}{16}$
2. Which is the largest: 0.024, 2.24, 2.224, or 0.424?
3. Round 0.7157 to the nearest hundredth.
4. Change 0.64 to a common fraction in lowest terms.
5. $0.3 + 0.6 + 0.2$ 6. $0.024 - 0.012$ 7. $5.6 + 30 + 0.003$
8. $\frac{1}{4} + 0.04 + 7.2$ 9. $31.05 - 1.273$ 10. Take 0.32 from 27.
11. Multiply 0.004 by 0.027   12. $0.2 \times 0.05 \times 3$
13. $4)\overline{0.0016}$ 14. Divide 35 by 0.5

DECIMAL TEST C
(see page 119.)

1. Write decimal equivalents for the following.
   (a) nine tenths   (b) seventy-two thousandths   (c) $\frac{7}{16}$
2. Which is largest: 1.6, 0.016, 0.066, 1.66, or 0.766?
3. Round 0.6142 to the nearest hundredth.
4. Change 0.65 to a common fraction in lowest terms.
5. $0.4 + 0.3 + 0.3$ 6. $0.018 - 0.009$ 7. $6.2 + 20 + 0.005$
8. $\frac{1}{4} + 0.06 + 7$ 9. $15.03 - 7.684$ 10. Take 0.16 from 30
11. Multiply 0.025 by 0.012   12. $0.4 \times 0.02 \times 6$
13. $6)\overline{0.003}$ 14. Divide 28 by 0.04

For additional practice with decimals see pages 213 to 218.

CHAPTER FOUR:
PERCENT, RATIO, AND PROPORTION

Percent is a special form of ratio (or fraction) in which the denominator is always 100. Because of its familiar and standardized format, a percent fraction is more meaningful to the average person than a comparable common fraction which may use any number for either its numerator or denominator.

ITEMS IN CHAPTER

Percent is particularly useful when a comparison is to be made. As you examine the tables below, you will see that although the same information is given in each, the second table in the form of percents makes a comparison that is easier to understand.

EXAMPLE

Four students did two sets of math tests one week apart. Use the tables below to assess the relative competence and progress of the four students. Table A gives raw scores; Table B shows the same scores converted to percents.

TABLE A

Student	A	B	C	D
First Test	30 out of 45	29 out of 40	15 out of 28	23 out of 35
Second Test	47 out of 55	38 out of 50	23 out of 35	48 out of 60

TABLE B

Student	A	B	C	D
First Test	66.7%	72.5%	53.6%	65.7%
Second Test	85.4%	76.0%	65.7%	80.0%

In spite of the preceding comments and in spite of the fact that everyone uses percent in day-to-day living, this topic is, for most students, the most difficult part of basic arithmetic. You will need to be especially careful as you work through the chapter.

ITEM P–1 CONVERSION: PERCENT TO FRACTION

STEPS

One: Replace the percent sign with the denominator 100.
Two: Simplify* in whatever way is necessary.

EXAMPLE

Change to common fractions: (a) 3%   (b) 27%

STEPS

One: $3\% = \dfrac{3}{100}$ $27\% = \dfrac{27}{100}$

Two: Not necessary since the fractions are already in lowest terms.

*In the following examples you will see that "simplify" can be interpreted in many ways.

EXAMPLES

Change to common fractions:
15%, 25%, 127%, 709%, 0.9%, $6\frac{1}{4}$%, 120%, 18.75%.

*Simplify may mean reduce to lowest terms.

$15\% = \dfrac{15}{100}$, which reduces to $\dfrac{3}{20}$. $25\% = \dfrac{25}{100}$, which reduces to $\dfrac{1}{4}$.

*Simplify may mean changing an improper fraction to a mixed number.

$127\% = \dfrac{127}{100}$ or $1\dfrac{27}{100}$ $709\% = \dfrac{709}{100}$ or $7\dfrac{9}{100}$

*Simplify may mean getting rid of a decimal point.

$0.9\% = \dfrac{0.9}{100}$, which is equal to $\dfrac{0.9 \times 10}{100 \times 10}$ or $\dfrac{9}{1000}$.

*Simplify may mean divide.

$6\frac{1}{4}\% = \dfrac{6\frac{1}{4}}{100}$, which is the same as $6\frac{1}{4} \div 100$ or $\dfrac{1}{16}$

*Simplify may mean any combination of these.

$120\% = \dfrac{120}{100} = 1\dfrac{20}{100} = 1\dfrac{1}{5}$ (Change mixed number, then reduce.)

$18.75\% = \dfrac{18.75}{100} = \dfrac{1875}{10\,000} = \dfrac{3}{16}$ (Get rid of decimal point, then reduce.)

These are just a few of the possibilities.

EXERCISE P – 1

Change the following percents to common fractions in lowest terms.

1. (a) 9% (b) 33% (c) 89% (These require *Step One* only.)

2. (a) 12% (b) 75% (c) 60% (d) 50% (These require reducing to lowest terms.)

3. (a) 150% (b) 275% (c) 320% (d) 102% (These are greater than 100%, so the fractional equivalent will be greater than 1 whole.)

4. (a) 0.4% (b) 1.2% (c) 12.5% (d) 0.95% (To get rid of a decimal point you use short-cut multiplication by a power of 10.)

5. (a) $37\frac{1}{2}$% (b) $8\frac{1}{3}$% (c) $1\frac{9}{10}$% (When the percent is a mixed number, you may use division by 100.)

6. (a) $\frac{1}{4}$% (b) $\frac{3}{8}$% (c) $\frac{2}{5}$% (d) $\frac{1}{10}$% (e) $\frac{7}{8}$% (f) $\frac{19}{20}$%
   (Since 1% $= \frac{1}{100}$, anything less than 1% will be less than $\frac{1}{100}$.)

7. (a) 85% (b) 106% (c) $16\frac{2}{3}$% (d) 0.5% (e) 2.25%
   (f) 700% (g) 28% (h) $\frac{1}{3}$% (i) $\frac{4}{5}$%

ANSWERS FOR EXERCISE P – 1

1. (a) 9% $= \frac{9}{100}$ (b) 33% $= \frac{33}{100}$ (c) 89% $= \frac{89}{100}$

2. (a) 12% $= \frac{12}{100} = \frac{3}{25}$ (b) 75% $= \frac{75}{100} = \frac{3}{4}$ (c) 60% $= \frac{60}{100} = \frac{3}{5}$
   (d) 50% $= \frac{50}{100} = \frac{1}{2}$

3. (a) 150% $= 1\frac{50}{100} = 1\frac{1}{2}$ (b) 275% $= 2\frac{75}{100} = 2\frac{3}{4}$ (c) 320% $= 3\frac{20}{100} = 3\frac{1}{5}$
   (d) 102% $= \frac{102}{100} = 1\frac{2}{100} = 1\frac{1}{50}$

4. (a) 0.4% $= \frac{0.4}{100} = \frac{4}{1000} = \frac{1}{250}$ (b) 1.2% $= \frac{1.2}{100} = \frac{12}{1000} = \frac{3}{250}$
   (c) 12.5% $= \frac{12.5}{100} = \frac{125}{1000} = \frac{1}{8}$ (d) 0.95% $= \frac{0.95}{100} = \frac{95}{10\,000} = \frac{19}{2000}$

5. (a) $37\frac{1}{2}$% $= 37\frac{1}{2} \div 100 = \frac{75}{2} \times \frac{1}{\underset{4}{100}} = \frac{3}{8}$ (b) $8\frac{1}{3}$% $= \frac{25}{3} \times \frac{1}{\underset{4}{100}} = \frac{1}{12}$

   (c) $1\frac{9}{10}$% $= \frac{1.9}{100} = \frac{19}{1000}$

6. (a) $\frac{1}{400}$ (b) $\frac{3}{800}$ (c) $\frac{1}{250}$ (d) $\frac{1}{1000}$ (e) $\frac{7}{800}$ (f) $\frac{19}{2000}$

7. (a) 85% $= \frac{17}{20}$ (b) 106% $= 1\frac{3}{50}$ (c) $16\frac{2}{3}$% $= \frac{1}{6}$ (d) 0.5% $= \frac{1}{200}$
   (e) 2.25% $= \frac{9}{400}$ (f) 700% $= 7$ (g) 28% $= \frac{7}{25}$ (h) $\frac{1}{3}$% $= \frac{1}{300}$
   (i) $\frac{4}{5}$% $= \frac{1}{125}$

ITEM P–2 CONVERSION: PERCENT TO DECIMAL

STEPS

One: Drop the percent sign.

Two: Move the decimal point two places left.

EXAMPLES

| 4.2% = 0.042 | 0.6% = 0.006 | 1.25% = 0.0125 | 187.5% = 1.875 |

If there is no written decimal point, it is understood to be after the final digit.

If 45% had a decimal point, it would be after the 5.   (45.0%)
If 180% had a decimal point, it would be after the 0.   (180.0%)
If 7% had a decimal point, it would be after the 7.   (7.0%)

Then, when you move the decimal point two places left, 45% becomes 0.45, 180% becomes 1.80 or 1.8, and 7% becomes 0.07.

If you have a fractional or mixed number percent, it is better to change it to a decimal percent before changing it to a decimal.

| $3\frac{1}{2}\% = 3.5\% = 0.035$ | $15\frac{1}{4}\% = 15.25\% = 0.1525$ |
| $\frac{7}{8}\% = 0.875\% = 0.00875$ | $162\frac{1}{2}\% = 162.5\% = 1.625$ |

EXERCISE P – 2

Change each of the following percents to a decimal equivalent.

1. (a) 3%   (b) 600%   (c) 0.4%   (d) 13%   (e) 0.75%
   (f) 18.75%   (g) 600.4%   (h) 6.52%
2. (a) $5\frac{1}{2}\%$ (b) $37\frac{1}{2}\%$ (c) $\frac{2}{5}\%$ (d) $206\frac{1}{4}\%$ (e) $12\frac{1}{2}\%$
   (f) $1\frac{1}{8}\%$ (g) $6\frac{1}{5}\%$ (h) $\frac{9}{20}\%$ (i) 1000%

ANSWERS FOR EXERCISE P – 2

1. (a) 3% = 0.03   (b) 600% = 6   (c) 0.4% = 0.004   (d) 13% = 0.13
   (e) 0.75% = 0.0075   (f) 18.75% = 0.1875   (g) 600.4% = 6.004
   (h) 6.52% = 0.0652
2. (a) $5\frac{1}{2}\% = 0.055$ (b) $37\frac{1}{2}\% = 0.375$ (c) $\frac{2}{5}\% = 0.004$ (d) $206\frac{1}{4}\% = 2.0625$
   (e) $12\frac{1}{2}\% = 0.125$ (f) $1\frac{1}{8}\% = 0.011\,25$ (g) $6\frac{1}{5}\% = 0.062$ (h) $\frac{9}{20}\% = 0.0045$
   (i) 1000% = 10

ITEM P-3 CONVERSION:
FRACTION OR DECIMAL TO PERCENT

STEPS

One: Multiply the decimal or fraction by 100.

Two: Affix the percent sign.

EXAMPLES

Change the following to percent. (a) $\frac{1}{4}$ (b) $6\frac{1}{2}$ (c) 0.2 (d) 1.5	
(a) $\frac{1}{4} = (\frac{1}{4} \times 100)\% = 25\%$	(b) $6\frac{1}{2} = (\frac{13}{2} \times \frac{100}{1})\% = 650\%$
(c) $0.2 = (0.2 \times 100)\% = 20\%$	(d) $1.5 = (1.5 \times 100)\% = 150\%$

A Common Sense Check
1 whole is equal to 100%
Anything less than 1 whole will be less than 100%.
Anything greater than 1 whole will be greater than 100%.

EXERCISE P – 3

1. Change the following fractions or decimals to percent.
   (a) $\frac{2}{5}$ (b) $\frac{1}{3}$ (c) 0.225   (d) $\frac{1}{8}$
   (e) $\frac{3}{10}$ (f) $\frac{7}{16}$ (g) 0.002   (h) 0.5
   (i) 0.75   (j) $\frac{7}{12}$ (k) $\frac{19}{20}$ (l) 0.01

2. Change the following whole or mixed numbers to percent.
   (a) 1   (b) 2   (c) $2\frac{1}{2}$ (d) 4.5   (e) 1.025
   (f) $3\frac{3}{5}$ (g) 3.2   (h) $1\frac{1}{3}$ (i) $2\frac{5}{16}$ (j) 2.75
   (k) 3.62   (l) 25   (m) $1\frac{1}{40}$

3. Check the following.

   Is each of your answers for Question 1 less than 100%?
   Is each of your answers for Question 2 greater than 100%?

4. Separate the following into two groups: less than 100% and greater than 100%.

   1%   3 $4\frac{1}{2}$ 86%   0.2   0.2%   2   20% $1\frac{1}{4}$ 1.2
   15%   175%   1.025

EXERCISE P – 3 (cont.)

5. Complete the following. Refer to the conversion chart if necessary.

	Common fraction	Decimal fraction	Percent
(a)	$\frac{1}{2}$		
(c)		0.025	
(e)			$4\frac{1}{4}\%$
(g)	$\frac{9}{16}$		

	Common fraction	Decimal fraction	Percent
(b)			255%
(d)		0.003	
(f)	$\frac{3}{400}$		
(h)		0.625	

SUMMARY: FRACTION, DECIMAL, PERCENT EQUIVALENTS

The chart below shows the six possible conversions.

Type of conversion	Method of conversion	Example of conversion
Fraction to decimal	Divide denominator into numerator.	$\frac{4}{5} = 5\overline{)4.0} = 0.8$
Decimal to fraction	Place numerator over appropriate denominator, and simplify.	$0.002 = \frac{2}{1000} = \frac{1}{500}$
Fraction to percent	Multiply by 100%.	$\frac{7}{8} = \frac{7}{8} \times 100\% = 87\frac{1}{2}\%$
Decimal to percent	Multiply by 100% (short method).	$0.035 = 0.035 \times 100\% = 3.5\%$
Percent to fraction	Replace the sign (%) with the denominator 100, and simplify.	$4.2\% = \frac{4.2}{100} = \frac{42}{1000} = \frac{21}{500}$
Percent to decimal	Drop the sign (%), and move the decimal point two places left.	$2.25\% = \frac{2.25}{100} = 0.0225$

ANSWERS FOR EXERCISE P - 3

1. (a) 40%   (b) $33\frac{1}{3}$%   (c) 22.5%   (d) 12.5%   (e) 30%
   (f) 43.75%   (g) 0.2%   (h) 50%   (i) 75%   (j) $58\frac{1}{3}$%
   (k) 95%   (l) 1%

2. (a) 100%   (b) 200%   (c) 250%   (d) 450%   (e) 102.5%
   (f) 360%   (g) 320%   (h) $133\frac{1}{3}$%   (i) 231.25%   (j) 275%
   (k) 362%   (l) 2500%   (m) 102.5%

4. Less than 100%: 1%, 86%, 0.2, 0.2%, 20%, 15%.
   More than 100%: 3, $4\frac{1}{2}$, 2, $1\frac{1}{4}$, 1.2, 175%, 1.025.

5. (a) $\frac{1}{2}$ 0.5   50%   (b) $2\frac{11}{20}$ 2.55   255%   (c) $\frac{1}{40}$ 0.025   2.5%
   (d) $\frac{3}{1000}$ 0.003   0.3%   (e) $\frac{17}{400}$ 0.0425 $4\frac{1}{4}$%   (f) $\frac{3}{400}$ 0.0075 $\frac{3}{4}$%
   (g) $\frac{9}{16}$ 0.5625 $56\frac{1}{4}$%   (h) $\frac{5}{8}$ 0.625   62.5%

Although any percent can be expressed in its equivalent common or decimal fraction form, it should be recognized that the function of percent is different from that of the other two types of fraction.

There are three possible "types" of problem in percent, each of them involving three different elements. These elements are: the two amounts that are related in a particular way, and the percent that defines that relationship. Examples of the three types are given below, and the rest of the chapter will be used to examine them in some detail.

Type One: Find 60% of 700   Answer: 420
Type Two: What percent is 420 of 700?   Answer: 60%
Type Three: If 60% of a number is 420, what is the number?   Answer: 700

Notice that the same three elements appear in each problem, two of them in the question and the other in the answer. All the work that you do in percent will fit into one of these "types."

ITEM P–4 FINDING A PERCENT OF A NUMBER

There are three commonly used methods of finding a percent of a number. It is good to know all three and be able to select the best method for a particular question. In actual practice, though, most people prefer one method and use it exclusively. You may wish to concentrate on one of the following methods and ignore the others. The three methods described here are:

The one-percent method

The fraction method

The decimal method

I The one-percent method

STEPS

One*: Find 1% of the number (divide by 100).

Two: Multiply by the given percent.

Example A

Find 42% of 658.
STEPS
One*: 1% of 658 is 6.58
Two: $42 \times 6.58 = 276.36$
42% of 658 is 276.36

Example B

Find 0.2% of 1400.
STEPS
One*: 1% of 1400 is 14
Two: $0.2 \times 14 = 2.8$
0.2% of 1400 is 2.8

*When dividing by 100, be sure to use the short-cut method which is to move the decimal point 2 places to the left.

$$658 \div 100 = 6.58. = 6.58 \qquad 1400 \div 100 = 14.00. = 14$$

Some other examples:

0.5% of $168 = 0.5 \times 1.68$ $= 0.84$	$2\frac{1}{2}\%$ of $160 = 2.5 \times 1.6$ $= 4$
120% of $1300 = 120 \times 13$ $= 1560$	800% of $240 = 800 \times 2.4$ $= 1920$

EXERCISE P – 4 (PART A)

1. Calculate the following using the one-percent method.

(a) 3% of 250   (b) 25% of 6   (c) 0.9% of 1850

(d) 155% of 200   (e) 3.2% of 70.8   (f) 0.2% of 420

(g) $\frac{1}{2}$% of 18   (h) $10\frac{3}{4}$% of 2800

ANSWERS

1. (a) $3 \times 2.5 = 7.5$ (b) $25 \times 0.06 = 1.5$ (c) $0.9 \times 18.5 = 16.65$

    (d) $155 \times 2 = 310$ (e) $3.2 \times 0.708 = 2.2656$ (f) $0.2 \times 4.2 = 0.84$

    (g) $0.5 \times 0.18 = 0.09$ (h) $10.75 \times 28 = 301$

II The fraction method

STEPS

One: Change the percent to a fraction.

Two: Multiply the fraction by the given number.

Example A

Find 55% of 600.

STEPS

One: $55\% = \frac{55}{100}$ or $\frac{11}{20}$

Two: $\frac{11}{20} \times \frac{600}{1} = 330$

55% of $600 = 330$

Example B

Find $6\frac{1}{4}$% of 20.

STEPS

One: $6\frac{1}{4}\% = \frac{25}{400}$ or $\frac{1}{16}$

Two: $\frac{1}{16} \times \frac{20}{1} = 1\frac{1}{4}$

$6\frac{1}{4}\%$ of $20 = 1\frac{1}{4}$

EXERCISE P – 4 (PART B)

1. Calculate the following:

(a) 10% of 600   (b) 150% of 32   (c) $12\frac{1}{2}$% of 160

(d) $33\frac{1}{3}$% of 3000   (e) 50% of 18.6   (f) $8\frac{1}{3}$% of 900

(g) 1.5% of 200   (h) 200% of 72.4

ANSWERS

1. (a) $\frac{1}{10} \times \frac{600}{1} = 60$ (b) $\frac{3}{2} \times \frac{32}{1} = 48$ (c) $\frac{1}{8} \times \frac{160}{1} = 20$

    (d) $\frac{1}{3} \times \frac{3000}{1} = 1000$ (e) $\frac{1}{2} \times 18.6 = 9.3$ (f) $\frac{1}{12} \times \frac{900}{1} = 75$

    (g) $\frac{3}{200} \times \frac{200}{1} = 3$ (h) $2 \times 72.4 = 144.8$

III The decimal method

STEPS

One: *Change the percent to a decimal.*

Two: *Multiply the decimal by the given number.*

Example A

Find 0.4% of 3000.

STEPS

One: 0.4% = 0.004

Two: 0.004 × 3000 = 12

0.4% of 3000 = 12

Example B

Find 225% of 148.2

STEPS

One: 225% = 2.25

Two: 2.25 × 148.2 = 333.45

225% of 148.2 = 333.45

EXERCISE P – 4 (PART C)

1. Calculate the following:

(a) 8% of 230   (b) 200% of 425   (c) 0.25% of 400

(d) 1.5% of 954   (e) 90% of 241   (f) 0.8% of 654

(g) $3\frac{1}{2}$% of 65.2   (h) 6% of 12.84

ANSWERS

1. (a) 0.08 × 230 = 18.4   (b) 2 × 425 = 850   (c) 0.0025 × 400 = 1

(d) 0.015 × 954 = 14.31   (e) 0.9 × 241 = 216.9   (f) 0.008 × 654 = 5.232

(g) 0.035 × 65.2 = 2.282   (h) 0.06 × 12.84 = 0.7704

ITEM P–5 FINDING WHAT PERCENT
ONE NUMBER IS OF ANOTHER

STEPS

One: *Form a fraction of the two numbers.*

Two: *Convert the fraction to a percent.*

Example A

You get 24 answers correct on a test consisting of 32 questions. What percent do you get correct?

STEPS

One: The fraction is $\dfrac{24}{32}$

Two: $\dfrac{24}{32}$ changed to percent is 75%

You get 75% of the questions correct.

Sometimes you must analyse the problem in order to find the numbers that form the fraction.

Example B

There are 25 men and 15 women in a class. What percent of the class is men?

STEPS

One: The fraction is $\dfrac{25}{40}$* $\left(\dfrac{\text{Number of men}}{\text{Total in class}}\right)$

Two: $\dfrac{25}{40}$ changed to a percent is $62\frac{1}{2}\%$

$62\frac{1}{2}\%$ of the class is men.

*Notice that the number 40 does not appear in the problem.

EXERCISE P – 5

1. Express the first number in each pair as a percent of the second.

   (a) 15 and 20   (b) 27 and 50   (c) 5.5 and 22   (d) 80 and 40

   (e) 2.5 and 10   (f) 162 and 540   (g) 75 and 60   (h) $\frac{3}{4}$ and $\frac{9}{16}$

2. A car which was purchased for $4000.00 sold at the end of the year for $3200.00. Find

   (a) the amount of depreciation

   (b) the depreciation as a percent of the original price

3. A house sold for $65 000.00. The agent's commission was $4550.00. What was the rate of commission?

4. A family spent $8320 of their total income for mortgage payments and taxes. If the total income was $32 000, what percent was used for this purpose?

5. A mixture has 2 parts of concentrate to 9 parts of water. What percent of the mixture is concentrate?

6. On the first day Elaine had 41 correct answers out of 62 questions on a quiz. On the following day her score was 36 out of a possible 55. On which day did Elaine get the higher percent?

7. On Monday I walked a mile in 15 minutes. By Friday I was able to walk the same distance in 12 minutes. By what percent had I decreased the time required to walk one mile?

8. The interest on $1800.00 is $193.50. What is the rate of interest?

ANSWERS FOR EXERCISE P -- 5

1. (a) $\frac{15}{20} \times 100\% = 75\%$ (15 is 75% of 20)   (b) 27 is 54% of 50

   (c) $\frac{5.5}{22} = \frac{55}{220} = \frac{1}{4} = 25\%$ (5.5 is 25% of 22)   (d) 80 is 200% of 40

   (e) $\frac{2.5}{10} = \frac{25}{100} = 25\%$ (2.5 is 25% of 10)   (f) 162 is 30% of 540

   (g) $\frac{75}{60} \times 100\% = 125\%$ (75 is 125% of 60)   (h) $\frac{3}{4} \div \frac{9}{16} \times 100\% = 133\frac{1}{3}\%$

2. (a) The amount of depreciation was $800.00. (4000 − 3200)

   (b) The percent of depreciation was 20%. ($\frac{\$800}{\$4000} \times 100\%$)

3. The rate of commission was 7%. ($\frac{\$4550}{\$65\ 000} \times 100\%$)

4. 26% of the total income was used for mortgage payments and taxes.

5. 18.18% is concentrate. ($\frac{2}{11} \times 100\%$)

6. Her score on the first day was higher. (41 out of 62 is 66.13%

   36 out of 55 is 65.45%)

7. I had decreased the time by 20%. (3 min is $\frac{1}{5}$ or 20% of 15 min.)

8. The rate of interest is 10.75%. (193.50 ÷ 1800 × 100%)

ITEM P–6 FINDING THE WHOLE (100%)
WHEN A GIVEN PERCENT IS KNOWN

There are two commonly used solutions; both are shown here.

Solution A	Solution B
STEPS	*STEPS*
One: Write the value of the given percent.	One: Form a fraction: $\dfrac{100\%}{given\,percent}$.
Two: Find the value of 1%.	Two: Multiply that fraction by the given number.
Three: Find the value of 100%.	

Example A

Interest on a short-term investment was 15%. If the interest was $1 200.00, what amount was invested?

STEPS	*STEPS*
One: If 15% = 1 200,	One: The ratio* is $\dfrac{100\%}{15\%}$
Two: then 1% $= \dfrac{1200}{15}$	Two: $\dfrac{100}{15} \times 1200 = 8000$
Three: and 100% $= \dfrac{1200}{15} \times 100,$	*ratio or fraction
which is equal to 8000.	

The amount invested was $8000.00.

Example B

Mrs. Glenn is on a low-calorie diet. At breakfast she consumed 25% of her total calories for the day, and at lunch she consumed another 30%. If she still had 540 calories left for the evening meal, what was her total calorie allowance for the day?

STEPS	*STEPS*
One: If 45% = 540,	One: The fraction is $\dfrac{100\%}{45\%}$
Two: then 1% = 540 ÷ 45	Two: $\dfrac{100}{45} \times 540 = 1200$
Three: and 100% $= \dfrac{540}{45} \times 100,$	
which equals 1 200.	

Her total calorie allowance for the day was 1 200.

EXERCISE P – 6

1. Complete the following.
    (a) 24% of _____ = 192   (b) 60% of _____ = 570
    (c) 2.5% of _____ = 2   (d) 4.25% of _____ = 76.5
    (e) 12.5% of _____ = 25   (f) $\frac{1}{4}$% of _____ = 70

2. (a) If 8% of a number is 96, what is the number?
   (b) If $1\frac{1}{4}$% of a number is 25, what is the number?
   (c) If 120% of a number is 30, what is the number?

Note: Some prefer an algebraic solution for this type of problem.
Solutions for the first question would begin like this:
(a) $0.24n = 192$;   (b) $0.6n = 570$;   (c) $0.025n = 2$;   etc.
The algebraic method will not be treated in detail in this chapter, but if you are familiar with it, by all means use it.

3. Three people each contributed 5% of their annual incomes for charitable purposes. If the contributions were respectively (a) $700.00, (b) $625.00, and
   (c) $1150.00, what were the corresponding incomes?

4. A radio was sold at a 20% discount. If the reduced price was $48.00, what was
   the original price?

5. During a chess tournament a player won 18 games, which was 90% of the total
   number of games that he played. How many games did he play?

6. If a 7% sales tax on a tape recorder is $8.40, what is the price of the tape
   recorder?

ANSWERS FOR EXERCISE P – 6

1. (a) 800   (b) 950   (c) 80   (d) 1800   (e) 200   (f) 28 000
2. (a) 1200   (b) 2000   (c) 25
3. (a) $14 000   (b) $12 500   (c) $23 000
4. The original price was $60.00. (80% = 48; 100% = 60)
5. He played 20 games. (90% = 18; 100% = 20)
6. The price of the recorder is $120.00 (7% = 8.40; 100% = 120)

ITEM P–7 THE THREE TYPES OF PROBLEM IN PERCENT

Any problem in percent involves three different elements. Two of these must be known in order to calculate the third.

The three elements are:

> The whole amount (100%)
>
> The "other" amount (may be more or less than the whole)
>
> The percent that the one amount is of the other.

Here are the three types of problems, showing the known elements and the one to be calculated.

TYPE A: To find a percent of a number   HOW TO DO IT

Find 40% of 500.		$40 \times 5 = 200$

We know:
  the percent (40%)
  the whole amount (500)
	We must find:
	  the other amount
		or $0.4 \times 500 = 200$
		or $\dfrac{2}{5} \times 500 = 200$

40% of 500 is 200.	200 is the "other" amount.	

TYPE B: To find what percent one number is of another   HOW TO DO IT

What percent of 500 is 200?		Form fraction: $\dfrac{200}{500}$

We know:
  the partial amount (200)
  the whole amount (500)
	We must find:
	  the percent
		Change to percent.
		$\dfrac{200}{500} \times 100\% = 40\%$

200 is 40% of 500.	40% is the answer.	

TYPE C: To find the whole if we know a part   HOW TO DO IT

If 40% of a number is 200, what is the number?		Divide 200 by 0.4

We know:
  the partial amount (200)
  the percent (40%)
	We must find:
	  the whole amount
		or divide 200 by $\dfrac{2}{5}$
		or $\dfrac{100\%}{40\%} \times 200 = 500$

If 40% of a number is 200, the number is 500.		

Some prefer to solve all percent problems using the proportion method. (See page 112).

Proportion equations for the preceding three examples are:

TYPE A	TYPE B	TYPE C
$40:100 = n:500$	$200:500 = p:100$	$40:100 = 200:w$
$n = \dfrac{40 \times 500}{100}$	$p = \dfrac{200 \times 100}{500}$	$w = \dfrac{100 \times 200}{40}$
$n = 200$	$p = 40$	$w = 500$

Solutions using common fractions:

$\dfrac{40}{100} \times 500 = 200$	$\dfrac{200}{500} \times 100\% = 40\%$	$\dfrac{100}{40} \times 200 = 500$

ITEM P-8 COMMON ERRORS IN PERCENT

1. INCORRECT USE OF THE PERCENT SIGN

This error takes two forms: either omitting the percent sign when it should be used, or using it when the answer is not a percent. There are three "types" of problem in percent. Only one of these requires an answer in percent. The three types are summarized and illustrated below.

Type One: To find a percent of a number
Example: Find 80% of 400.
Answer: 80% of 400 is <u>320</u>.   Answer is *not* in percent.

Type Two: To find what percent one number is of another
Example: What percent of 400 is 320?
Answer: 320 is <u>80%</u> of 400.   Answer *is* in percent.

Type Three: To find 100% if a given percent is known
Example: If 80% of a number is 320, what is the number?
Answer: The number is <u>400</u>.   Answer is *not* in percent.

Notice in the calculations below that in *Types One* and *Three*, the percent signs are "cancelled out."

Type One	Type Two	Type Three
$\dfrac{80\%}{100\%} \times 400 = 320$	$\dfrac{320}{400} \times 100\% = 80\%$	$\dfrac{100\%}{80\%} \times 320 = 400$

2. CONFUSING A FRACTIONAL PERCENT WITH A FRACTION OF THE SAME NAME

   Example: $\frac{1}{4}$% is *not* equal to $\frac{1}{4}$; instead it is $\frac{1}{4}$ of 1%, or $\frac{1}{400}$.

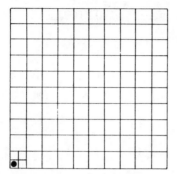

This diagram shows $\frac{1}{4}$%.

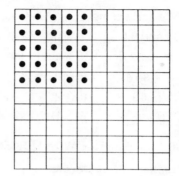

This diagram shows $\frac{1}{4}$ (or 25%).

PRACTICE WITH PERCENT (I)

1. Complete the chart to show equivalents.

Fraction	Decimal	Percent
	0.175	
$\frac{1}{400}$		
		2.5%
	0.833	
		$\frac{1}{10}$%
$\frac{9}{16}$		
	0.005	
		187.5%
	1.55	
$\frac{11}{15}$		

2. Find the total of each set.

   (a) 6% of 1400
        1% of 300
        0.5% of 1200
        150% of 36
        $+\frac{1}{4}$% of 3600
        _____

   (b) 2.5% of 200
        20% of 450
        0.125% of 16 000
        $83\frac{1}{3}$% of 300
        $+0.9$% of 1500
        _____

   (c) 187.5% of 200
        1.4% of 500
        0.3% of 1680
        62.5% of 1200
        $+90$% of 400
        _____

3. Express the first number in each pair as a percent of the second.
   (a) 20 and 50
   (b) 0.5 and 1.5
   (c) 16 and 160
   (d) $\frac{1}{2}$ and 1
   (e) 25 and 45
   (f) 1 day and 1 week
   (g) 50¢ and 1 dime
   (h) 18 months & 1 year
   (i) 0.03 and 0.24
   (j) 95 and 9.5
   (k) 72 in. and 1 yd.

4. In each case find 100% using the information given.
   (a) 40% is 14
   (b) 2.5% is 100
   (c) 150% is 36
   (d) 75% is 21
   (e) 9% is 45
   (f) 20% is 1.2
   (g) 0.5% is 14
   (h) 60% is 900
   (i) 225% is 72
   (j) 1000% is 400

ANSWERS FOR PRACTICE WITH PERCENT (I)

1.

Fraction	Decimal	Percent
$\frac{7}{40}$	0.175	17.5%
$\frac{1}{400}$	0.0025	0.25%
$\frac{1}{40}$	0.025	2.5%
$\frac{5}{6}$	0.833	$83\frac{1}{3}$%
$\frac{1}{1000}$	0.001	0.1%
$\frac{9}{16}$	0.5625	56.25%
$\frac{1}{200}$	0.005	0.5%
$1\frac{7}{8}$	1.875	187.5%
$1\frac{11}{20}$	1.55	155%
$\frac{11}{15}$	0.733	$73\frac{1}{3}$%

2. (a)   84   (b)   5   (c)   375
          3   90   7
          6   20   5.04
          54   250   750
          +9   +13.5   +360
         ─────   ──────   ───────
          156   378.5   1497.04

3. (a) 40%   (b) $33\frac{1}{3}$%   (c) 10%   (d) 50%
   (e) $55\frac{5}{9}$%   (f) $14\frac{2}{7}$%   (g) 500%   (h) 150%
   (i) 12.5%   (j) 1000%   (k) 200%

4. (a) 35   (b) 4000   (c) 24   (d) 28   (e) 500
   (f) 6   (g) 2800   (h) 1500   (i) 32   (j) 40

PRACTICE WITH PERCENT (II)

1. Write the fraction and decimal equivalents for each of the following:
   - (a) 30%
   - (b) 195%
   - (c) 0.2%
   - (d) 1.25%
   - (e) 87.5%
   - (f) 110.5%
   - (g) 0.01%
   - (h) $33\frac{1}{3}\%$
   - (i) $8\frac{1}{3}\%$
   - (j) $66\frac{2}{3}\%$
   - (k) $6\frac{1}{4}\%$
   - (l) $18\frac{3}{4}\%$
   - (m) $12\frac{1}{2}\%$
   - (n) 450%

2. Express each of the following as a percent.
   - (a) $\frac{3}{4}$
   - (b) $\frac{11}{16}$
   - (c) $\frac{5}{12}$
   - (d) $2\frac{1}{2}$
   - (e) $5\frac{5}{8}$
   - (f) $\frac{1}{100}$
   - (g) $1\frac{3}{40}$
   - (h) $\frac{17}{20}$

3. Change these decimals to percent equivalents.
   - (a) 0.4
   - (b) 0.55
   - (c) 0.5625
   - (d) 9.1
   - (e) 0.2275
   - (f) 0.0025

4. Find the total for each set.

(a)	(b)	(c)
6% of 4000	2% of 900	10% of 850
3.2% of 900	1.6% of 4600	7.5% of 2000
400% of 188	300% of 490	200% of 150
$\frac{1}{4}\%$ of 800	$\frac{1}{5}\%$ of 600	$\frac{7}{8}\%$ of 1200
+0.4% of 348	+0.2% of 560	+0.3% of 720

5. Express the first number in each pair as a percent of the second.
   - (a) 2 and 4
   - (b) 11 and 33
   - (c) 16 and 20
   - (d) 2.5 and 10
   - (e) 8 and 6
   - (f) 14 and 35
   - (g) 0.2 and 1.6
   - (h) 0.2 and 160
   - (i) 50 and 50
   - (j) 46 and 69
   - (k) 30 and 3000
   - (l) $\frac{3}{4}$ and $\frac{7}{8}$

6. Find the value of 100% in each case.
   - (a) 20% = 40
   - (b) 45% = 9
   - (c) 62.5% = 420
   - (d) 0.1% = 14
   - (e) 400% = 32
   - (f) $8\frac{1}{2}\%$ = 68
   - (g) $\frac{1}{4}\%$ = 5.2
   - (h) 225% = 0.81

7. Complete the following chart.

Percent	Partial amount	Whole or 100%
12%	60	
4%		300
	45	225
0.5%	25	
0.5%		25

Percent	Partial Amount	Whole or 100%
0.125%		1200
350%	21	
	0.25	0.6
$\frac{1}{2}\%$		2
	3000	300

ANSWERS FOR PRACTICE WITH PERCENT (II)

1. (a) $\frac{3}{10}$, 0.3   (b) $1\frac{19}{20}$, 1.95   (c) $\frac{1}{500}$, 0.002   (d) $\frac{1}{80}$, 0.0125
   (e) $\frac{7}{8}$, 0.875   (f) $1\frac{21}{200}$, 1.105   (g) $\frac{1}{10\,000}$, 0.0001   (h) $\frac{1}{3}$, 0.3$\dot{3}$
   (i) $\frac{1}{12}$, 0.08$\dot{3}$ (j) $\frac{2}{3}$, 0.6$\dot{6}$ or 0.667   (k) $\frac{1}{16}$, 0.0625
   (l) $\frac{3}{16}$, 0.1875   (m) $\frac{1}{8}$, 0.125   (n) $4\frac{1}{2}$, 4.5

2. (a) 75%   (b) 68.75% or $68\frac{3}{4}$%   (c) $41\frac{2}{3}$%   (d) 250%
   (e) 562.5%   (f) 1%   (g) 107.5%   (h) 85%

3. (a) 40%   (b) 55%   (c) 56.25%   (d) 910%   (e) 22.75%   (f) 0.25%

4. (a) 240   (b)   18   (c)   85
        28.8   73.6   150
        752   1470   300
        2   1.2   10.5
       +1.392   +1.12   +2.16
      ─────────   ─────────   ─────────
       1024.192   1563.92   547.66

5. (a) 50%   (b) $33\frac{1}{3}$%   (c) 80%   (d) 25%
   (e) $133\frac{1}{3}$%   (f) 40%   (g) $12\frac{1}{2}$%   (h) $\frac{1}{8}$%
   (i) 100%   (j) $66\frac{2}{3}$%   (k) 1%   (l) $85\frac{5}{7}$%

6. (a) 200   (b) 20   (c) 672   (d) 14 000
   (e) 8   (f) 800   (g) 2080   (h) 0.36

7.

		500	
	12		
20%			
		5000	
	0.125		

	1.5	
		6
$41\frac{2}{3}$%		
	0.01	
1000%		

For additional practice with percent see pages 219 and 220.

PERCENT TESTS
(see page 119.)

Note: Answers are considered wrong if a necessary percent sign is omitted, or if a percent sign is used incorrectly. Example: 3% of 5000 is 150 (not 150%).

PERCENT TEST A

1. Express as common fractions in lowest terms.
   (a) 5%   (b) $3\frac{1}{2}$%   (c) $18\frac{3}{4}$%   (d) 4.6%
2. Change to percent form.   (a) $\frac{1}{4}$ (b) $3\frac{1}{2}$ (c) 0.024
3. Find each result.   (a) 60% of 400   (b) 4% of 985   (c) 0.3% of 1600
4. Solve.   (a) 16 is ____% of 40   (b) 1 ounce is ____% of 1 pound
5. What percent of a dollar is 63¢?   6. 80% of what number is 16?
7. If 6% of a number is 24, what is the number?
8. If 32% of a number is 640, what is the number?

PERCENT TEST B

1. Express as common fractions in lowest terms.
   (a) 5%   (b) $1\frac{1}{2}$%   (c) $31\frac{1}{4}$%   (d) 3.2%
2. Change to percent form.   (a) $\frac{3}{4}$ (b) $1\frac{1}{2}$ (c) 0.016
3. Find each result.   (a) 80% of 600   (b) 2% of 345   (c) 0.2% of 1400
4. (a) 14 is ____% of 20.   (b) 1 minute is ____% of 1 hour.
5. What percent of a dollar is 35¢?
6. If 4% of a number is 16, what is the number?
7. If 60% of a number is 12, what is the number?
8. If 40% of a number is 240, what is the number?

PERCENT TEST C

1. Express as common fractions in lowest terms.
   (a) 15%   (b) $3\frac{1}{2}$%   (c) $31\frac{1}{4}$%   (d) 4.5%
2. Change to percent form.   (a) $\frac{1}{4}$ (b) $2\frac{1}{2}$ (c) 0.024
3. Find each result.   (a) 40% of 800   (b) 3% of 240   (c) 0.4% of 1300
4. Solve.   (a) 16 is ____% of 80.   (b) 1 in. is ____% of 1 ft.
5. What percent of a dollar is 37¢?   6. 5% of what number is 20?
7. If 30% of a number is 15, what is the number?
8. If 70% of a number is 350, what is the number?

ITEM P-9 RATIO AND PROPORTION:
MEANING AND TERMINOLOGY

Numbers may be compared by subtraction or by division.

EXAMPLE: Compare the numbers 10 and 30.

By subtraction	By division
10 is 20 less than 30	10 is $\dfrac{10}{30}$ or $\dfrac{1}{3}$ of 30
30 is 20 greater than 10	30 is $\dfrac{30}{10}$ or 3 times as great as 10

When two numbers are compared by division, the resulting fraction is called a ratio.

$$\text{The ratio of 10 to 30 is } \frac{1}{3} \text{ (read "one to three").}$$

$$\text{The ratio of 20 to 25 is } \frac{4}{5} \text{ (read "4 to 5").}$$

$$\text{The ratio of 60 to 75 is } \frac{4}{5}.$$

A *proportion* is a statement that two ratios are equal.

40 is related to 80 in the same way that 16 is related to 32.

This may be expressed as a proportion in three different forms.

$\dfrac{40}{80} = \dfrac{16}{32}$	40:80 = 16:32	40/80 = 16/32

A proportion has four terms: first, second, third, and fourth.

$$\begin{array}{ccccccc} 40 & : & 80 & = & 16 & : & 32 \\ \textit{1st} & & \textit{2nd} & & \textit{3rd} & & \textit{4th} \end{array} \quad \text{or} \quad \begin{array}{l} \text{first} \rightarrow \dfrac{40}{80} = \dfrac{16}{32} \begin{array}{l} \leftarrow \text{third} \\ \leftarrow \text{fourth} \end{array} \\ \text{second} \rightarrow \end{array}$$

The first and fourth terms are the outer pair, or the extremes.

The second and third terms are the inner pair, or the means.

$$\begin{array}{c} \overset{\frown}{\text{outer pair}} \\ 40:80 = 16:32 \\ \underset{\smile}{\text{inner pair}} \end{array}$$

In a proportion, the product of the outer pair is equal to the product of the inner pair.

In the proportion 40:80 = 16:32,

the product of the outer pair is 1 280 (40 × 32)

the product of the inner pair is 1 280.   (80 × 16)

If the proportion is written in the form of equivalent fractions, the cross products are equal.

$$\frac{40}{80} \diagdown\!\!\!\!\!\diagup \frac{16}{32}$$

ITEM P–10 RATIO AND PROPORTION:
FINDING THE MISSING TERM

If three terms of a proportion are known, the fourth can be calculated by applying the rule that the products of outer and inner pairs must be equal.

EXAMPLE

Find the missing term in the proportion 34:25 = 51:t.

The product of the outer pair is $34t$ ($34 \times t$).

The product of the inner pair is 1 275 (25×51).

Since this is a proportion, the two products are equal.

Therefore, $34t = 1275$

and $\quad t = \dfrac{1275}{34}$ or 37.5

Some other examples:

Find the missing term in each of the following proportions.

(a) t:2 = 7:3	(b) 6/16 = t/8	(c) $\dfrac{4}{5} = \dfrac{18}{t}$	(d) 2:t = 15:52.5
Outer pair: $3t$ Inner pair: 14 $3t = 14$ $t = 14 \div 3$ $t = 4\frac{2}{3}$ $4\frac{2}{3}$:2 = 7:3	Outer pair: 48 Inner pair: $16t$ $16t = 48$ $t = 48 \div 16$ $t = 3$ 6/16 = 3/8	Outer pair: $4t$ Inner pair: 90 $4t = 90$ $t = \dfrac{90}{4}$ $t = 22.5$ $\dfrac{4}{5} = \dfrac{18}{22.5}$	Outer pair: 105 Inner pair: $15t$ $15t = 105$ $t = \dfrac{105}{15}$ $t = 7$ 2:7 = 15:52.5

Or you may prefer this version:

To find the missing term in a proportion, divide the product of the opposite pair by the "partner" of the unknown term.

EXAMPLE

Find the value of *y* in each proportion.

(a) 50:20 = *y*:4   (b) *y*:36 = 15:60   (c) 0.5:*y* = 2.5:20

$$y = \frac{50 \times 4}{20}$$

$$y = 10$$

$$y = \frac{36 \times 15}{60}$$

$$y = 9$$

$$y = \frac{0.5 \times 20}{2.5}$$

$$y = 4$$

Then substitute the value of *y* in the proportion and prove.

$$\frac{50}{20} = \frac{10}{4}$$

Proof

50 × 4 = 200

20 × 10 = 200

$$\frac{9}{36} = \frac{15}{60}$$

Proof

9 × 60 = 540

36 × 15 = 540

$$\frac{0.5}{4} = \frac{2.5}{20}$$

Proof

0.5 × 20 = 10

4 × 2.5 = 10

EXERCISE P -- 9 and 10

1. Use the rule of cross multiplication to determine which of the following pairs of ratios form a proportion (in other words, which pairs are equal).

(a) 5/4, 30/24   (b) 10/16, 25/32   (c) 0.3/0.9, 30/90

(d) $\frac{1}{2}:\frac{3}{4}$ and $2\frac{1}{2}:3\frac{3}{4}$ (e) $\frac{1.2}{1.6}$ and $\frac{9.6}{12.8}$ (f) $\frac{16}{9}$ and $\frac{88}{50}$

(g) $\frac{42}{49}$ and $\frac{35}{42}$ (h) $\frac{200}{450}$ and $\frac{8}{18}$

2. Find the missing term in each of the following proportions.

(a) 20:60 = *a*:90   (b) 45/22.5 = 0.8/*h*   (c) $1\frac{1}{4}:15 = 2\frac{1}{2}:m$

(d) 5:85 = *h*:510   (e) 0.05:*d* = 2:60   (f) $z:\frac{3}{4} = 4:9$

(g) 0.5:0.1 = *k*:1.5   (h) *y*:20 = 50:25

3. Prove your answers for question 2 by substituting the value of the missing term, and then using cross multiplication.

EXAMPLE

The answer for (a) is 30.

The proportion now reads: 20:60 = 30:90

Proof: 20 × 90 = 1800   60 × 30 = 1800

4. (a) Two numbers have a ratio of 4:5. If the lesser number is 92, what is the other number?

EXAMPLE

The proportion is $4:5 = 92:n$

Therefore $4n = 92 \times 5$

$$n = \frac{92 \times 5}{4} \text{ or } 115$$

The other number is 115.

Proof: $4 \times 115 = 460 \qquad 5 \times 92 = 460$

(b) The ratio of two numbers is $7:3$. If the larger of the numbers is 56, what is the other number?

(c) A father's investment in a business and his son's are in the ratio $\frac{1}{2}:5$. If the father invests $80 000.00, what is the amount of the son's investment?

(d) I use my car for business and for pleasure. In an average month, the ratio of the former to the latter is $5:2$. If I drive 600 miles for business purposes, how many miles do I drive for pleasure?

(e) The income from an investment last year was in a ratio of $1\frac{1}{2}:10$ when compared with the amount of the investment. What percent was that?

(f) The ratio of powder to liquid in a solution is $2:9$. How many cups of powder must be added to 50 cups of liquid?

ANSWERS FOR EXERCISE P – 9 and 10

1. (a) $5 \times 24 = 120 \qquad 4 \times 30 = 120$ (equal)
   (b) $10 \times 32 = 320 \qquad 16 \times 25 = 400$ (not equal)
   (c) $0.3 \times 90 = 27 \qquad 0.9 \times 30 = 27$ (equal) (d) $\frac{1}{2} \times 3\frac{3}{4} = 1\frac{7}{8} \qquad \frac{3}{4} \times 2\frac{1}{2} = 1\frac{7}{8}$ (equal)
   (e) $1.2 \times 12.8 = 15.36 \qquad 1.6 \times 9.6 = 15.36$ (equal)
   (f) $16 \times 50 = 800 \qquad 9 \times 88 = 792$ (not equal)
   (g) $42 \times 42 = 1764 \qquad 49 \times 35 = 1715$ (not equal)
   (h) $200 \times 18 = 3600 \qquad 450 \times 8 = 3600$ (equal)

2. (a) $a = \dfrac{20 \times 90}{60} = 30$ (b) $h = \dfrac{22.5 \times 0.8}{45} = 0.4$ (c) $m = \dfrac{15 \times 2.5}{1.25} = 30$

   (d) $h = \dfrac{5 \times 510}{85} = 30$ (e) $d = \dfrac{0.05 \times 60}{2} = 1.5$ (f) $z = \dfrac{0.75 \times 4}{9} = \dfrac{1}{3}$

   (g) $k = \dfrac{0.5 \times 1.5}{0.1} = 7.5$ (h) $y = \dfrac{20 \times 50}{25} = 40$

4. (b) $7:3 = 56:n$ The number is 24. (c) $\dfrac{0.5}{5} = \dfrac{n}{80\ 000}$ The son's share is $8000.00.

   (d) $5:2 = 600:n$ I drive 240 miles for pleasure.
   (e) $1.5:10 = n:100$ (since percent must be expressed in hundredths). The income is 15%.
   (f) $2:9 = n:50 \quad 11\frac{1}{9}$ cups of powder must be added.

ITEM P–11   RATIO AND RATE

The terms of a proportion may be rearranged without affecting the relationship. 20 and 80 are related to each other in the same way that 24 and 96 are related. This proportion may be stated in four different ways:

20:80 = 24:96	80:20 = 96:24	20:24 = 80:96	24:20 = 96:80

Prove the truth of these proportions by comparing inner and outer products. Using the fractional form, you will see that fractions are equivalent.

$\dfrac{20}{80} = \dfrac{24}{96}$	$\dfrac{80}{20} = \dfrac{96}{24}$	$\dfrac{20}{24} = \dfrac{80}{96}$	$\dfrac{24}{20} = \dfrac{96}{80}$
Both reduce to $\frac{1}{4}$.	Both reduce to $\frac{4}{1}$.	Both reduce to $\frac{5}{6}$.	Both reduce to $\frac{6}{5}$.

The terms of a ratio must refer to the same unit of measurement. If different units of measurement are compared, the resulting fraction is called a "rate" rather than a ratio. If two rates are equal, they can be expressed as a proportion.

For example, a train travels 240 km in 2 h, or 600 km in 5 h. The proportion derived from these two rates is $\dfrac{240}{2} = \dfrac{600}{5}$.

In each case the rate is $\dfrac{\text{distance}}{\text{time}}$ (the units of measurement are different).

The same information may be stated as a ratio proportion, in which distance is compared with distance and time is compared with time.

$$\left(\dfrac{\text{distance}}{\text{distance}}\right) \quad \dfrac{240}{600} = \dfrac{2}{5} \quad \left(\dfrac{\text{time}}{\text{time}}\right)$$

Examine the proportions in the following examples. The first proportion for each problem is a rate proportion; the second is composed of ratios.

Example A

At a cost of \$4.18 for 8 articles, how many could you buy for \$10.45?

Rate Proportion:

$$\dfrac{\text{cost}}{\text{quantity}} = \dfrac{\text{cost}}{\text{quantity}} \qquad \dfrac{418}{8} = \dfrac{1045}{y} \qquad y = \dfrac{8 \times 1045}{418} = 20$$

Ratio Proportion:

$$\dfrac{\text{cost}}{\text{cost}} = \dfrac{\text{quantity}}{\text{quantity}} \qquad \dfrac{418}{1045} = \dfrac{8}{y} \qquad y = \dfrac{1045 \times 8}{418} = 20$$

You could buy 20 articles for \$10.45.

Although proportion problems may be solved using rates rather than ratios, it is an advantage to be aware of the difference and to be able to convert a rate problem to a ratio type if necessary. Some problems in proportion cannot be solved unless the units of measurement match. (This is true of inverse proportion).

EXERCISE P – 11

For each of the following problems, write and solve
    (a) a rate proportion   (b) a ratio proportion

1. At a cost of 95¢ for 7 articles, how many could you buy for $11.40?

2. At a rate of 362 km in 4 h, how long will it take a car to travel 1991 km?

3. A pieceworker in a factory can complete 25 items in $31\frac{1}{4}$ min. At that rate, how many items can she complete in 1 h?

4. A car uses 10 L of gasoline every 85 km. At that rate, how much gas will be used on a trip that covers 3300 km?

5. My watch needed repair. It lost 2 min 30 s every day. At that rate, how many seconds did it lose per hour?

6. If it cost $48.00 to paper a wall that is 5 m long and 3 m high, what would it cost to paper the other three walls that have a total surface area of 20 m²? (Window and door areas are excluded.)

ANSWERS FOR EXERCISE P – 11

1. (a) $\frac{95}{7} = \frac{1140}{n}$ (b) $\frac{95}{1140} = \frac{7}{n}$ $n = 84$ You could buy 84 articles.

2. (a) $\frac{362}{4} = \frac{1991}{n}$ (b) $\frac{362}{1991} = \frac{4}{n}$ $n = 22$ It will take 22 h.

3. (a) $\frac{25}{31\frac{1}{4}} = \frac{b}{60}$ (b) $\frac{25}{b} = \frac{31\frac{1}{4}}{60}$ $b = 48$ She can complete 48 items.

4. (a) $\frac{85}{10} = \frac{3300}{x}$ (b) $\frac{85}{3300} = \frac{10}{x}$ $x = 388.24$ 388.24 L will be used.

5. (a) $\frac{150}{24} = \frac{s}{1}$ (b) $\frac{1}{24} = \frac{s}{150}$ $s = 6.25$ It lost $6\frac{1}{4}$ s/h.

6. (a) $\frac{15}{48} = \frac{20}{m}$ (b) $\frac{15}{20} = \frac{48}{m}$ $m = 64$ It would cost $64.00.

ITEM P–12 PERCENT AS A RATIO

Any type of percent problem may be solved using the proportion method.

EXAMPLES OF CONVERSIONS: FRACTIONS OR DECIMALS TO PERCENT

Express the following as percents.		
(a) $\dfrac{3}{5}$	(b) 0.2	(c) $1\dfrac{1}{4}$
$3:5 = p:100$	$2:10 = p:100$	$5:4 = p:100$
$p = \dfrac{3 \times 100}{5}$	$p = \dfrac{2 \times 100}{10}$	$p = \dfrac{5 \times 100}{4}$
$p = 60$	$p = 20$	$p = 125$
Answer: 60%	Answer: 20%	Answer: 125%

Problem Type 1: Finding a percent of a number

EXAMPLES

Find the following:		
(a) 20% of 5000	(b) 0.3% of 2000	(c) 220% of 4
$20:100 = a:5000$	$0.3:100 = a:2000$	$220/100 = a/4$
$a = \dfrac{20 \times 5000}{100}$	$a = \dfrac{0.3 \times 2000}{100}$	$a = \dfrac{220 \times 4}{100}$
$a = 1000$	$a = 6$	$a = 8.8$
20% of 5000 = 1000	0.3% of 2000 = 6	220% of 4 = 8.8

Problem Type 2: Finding what percent one number is of another

EXAMPLES

Express the first number in each pair as a percent of the second.

(a) 40 and 50	(b) 2.6 and 52	(c) $\dfrac{1}{4}$ and $\dfrac{3}{8}$
$\dfrac{40}{50} = \dfrac{r}{100}$	$\dfrac{2.6}{52} = \dfrac{r}{100}$	$\dfrac{1}{4} : \dfrac{3}{8} = r:100$
$r = \dfrac{40 \times 100}{50}$	$r = \dfrac{2.6 \times 100}{52}$	$r = \dfrac{\frac{1}{4} \times 100}{\frac{3}{8}}$
$r = 80$	$r = 5$	$\frac{1}{4}$ is $66\frac{2}{3}\%$ of $\frac{3}{8}$
40 is 80% of 50	2.6 is 5% of 52	

Problem Type 3: Finding 100% if a given percent is known

EXAMPLES

Find the value of 100% if

(a) 32% = 512	(b) 0.5% = 0.4	(c) 275% = 159.5
$\dfrac{100}{32} = \dfrac{w}{512}$	$\dfrac{100}{0.5} = \dfrac{w}{0.4}$	$\dfrac{100}{275} = \dfrac{w}{159.5}$
$w = \dfrac{100 \times 512}{32}$	$w = \dfrac{100 \times 0.4}{0.5}$	$w = \dfrac{100 \times 159.5}{275}$
$w = 1600$	$w = 80$	$w = 58$
32% of 1600 = 512	0.5% of 80 = 0.4	275% of 58 = 159.5

POST TEST – PART ONE

WHOLE NUMBERS

W – 1 Give the place value of each underlined digit: $\underline{6}2$ $3\underline{9}5$ 684.

W – 2 Write 52 650 007 803 in words.

W – 3 Write numerals for the following: three billion, four hundred million, eight thousand, two hundred forty-one

W – 4 Show that the following problem can be done using subtraction.
565 divided by 113.

W – 5 (a) Find the average of 400, 456, 398, 424, and 412.

(b) Write the power composed of base 10, and exponent 3.

(c) Find the value of 15^3.

(d) What is the square root of 144?

(e) Find the value of $2(4 + 5 \times 3) \div \dfrac{24 + 36}{17 - 2}$

W – 6 (a) Add 44 587, 6390, 228 758, 29, 587, and 32 907.

(b) Subtract 56 873 from 100 000.

(c) Multiply 6071 by 4900.

(d) Divide 27 into 1 080 810.

W – 7 Calculate and prove each answer using the inverse operation.

(a) 62 000 − 20 078   (b) 460 × 350   (c) 72 395 ÷ 240

COMMON FRACTIONS

F – 1 (a) Write a fraction with numerator 10.

(b) Change $13\frac{5}{8}$ to an improper fraction.

F – 2 (a) Reduce $\frac{95}{133}$ to lowest terms.

(b) Choose the fractions that are equivalent to $\frac{7}{8}$:
$\frac{3}{4}$ $\frac{21}{32}$ $\frac{49}{56}$ $\frac{112}{128}$ $\frac{140}{156}$

F – 3 Add the following: $16\frac{1}{4} + 2\frac{3}{5} + 21 + 3\frac{5}{12} + \frac{9}{10}$

F – 4 Subtract $7\frac{11}{16}$ from $15\frac{1}{8}$.

F – 5 Multiply the following: $2\frac{2}{3} \times 24 \times 5\frac{5}{8} \times \frac{7}{9}$

F – 6 Divide (a) 20 by $\frac{4}{5}$ (b) $\frac{4}{5}$ by 20   (c) $2\frac{5}{8} \div 1\frac{1}{3}$

F – 7 What fraction is 45 of 75?

F – 8 If $\frac{7}{16}$ of a number is 56, what is the number?

DECIMALS

D – 1 Which of the following is equal to 25.0340?

(a) 25 and 34 hundredths (b) $25\frac{340}{1000}$

(c) twenty-five and 340 ten-thousandths

(d) 25 and 34 ten-thousandths

D – 2 Write numerals for the following:

(a) two hundred fifty-seven thousandths

(b) two hundred and fifty-seven thousandths

D – 3 Change 0.016 to a common fraction in lowest terms.

D – 4 Add or subtract as indicated.

(a) 0.65 + 0.35 (b) 50 + 18.6 + 0.59 + 2 + 61.03

(c) 20 − 0.55 (d) 47 hundredths − 47 thousandths

D – 5 Multiply.

(a) 0.3 by 0.1 by 0.2 (b) 4.6 by 0.0012

D – 6 Divide.

(a) 400 by 0.004 (b) $250)\overline{7.5}$

(c) $\frac{9}{11}$ (correct to three places)

D – 7 Complete the following:

(a) To multiply by 10, move the decimal point __ places to the _____.

(b) To divide by 10 000, move the decimal point __ places to the _____.

D – 8 Round (off) as indicated.

(a) 0.99 to the nearest tenth (b) 186.4539 to two places

(c) 1694.5575 to the nearest whole number

D – 9 Change to decimal form.

(a) $15\frac{6}{25}$ (b) $21\frac{1}{7}$ (c) $\frac{3}{32}$

D – 10 Solve using common fractions: $3\frac{1}{3} \times 0.75$

PERCENT

P – 1 Change each percent to a fraction in lowest terms:

(a) 65% (b) $\frac{1}{200}$% (c) 62.5% (d) 275%

P – 2 Change to decimal form.

(a) 463% (b) 0.05% (c) $20\frac{7}{8}$%

P – 3 Change to percent equivalents.

(a) 0.3125 (b) 4.03 (c) $\frac{1}{12}$ (d) $\frac{3}{40}$

P – 4 Calculate the results.

(a) 5% of 3200   (b) 450% of 16   (c) 6.025% of 260

(d) $\frac{9}{10}$% of 1 500

P – 5 Express the first number in each pair as a percent of the second.

(a) 25 and 55   (b) $\frac{2}{5}$ and 1.2   (c) 0.06 and 60

P – 6 In each case, find the value of 100% given the following information:

(a) 8% of the number is equal to 15   (b) 2.3% of the number is equal to 8.05

(c) 0.2% of the number is equal to 50

P – 7 In a proportion, the product of the _____ terms is equal to the product of the _____ terms.

P – 8 Find the value of *t* in the proportion 24:90 = 36:*t*.

P – 9 Write a rate proportion and a ratio proportion to show the relationship between 25 articles for 95¢ and 15 for 57¢.

ANSWERS FOR POST TEST – PART ONE

WHOLE NUMBERS

W – 1 6 is tens of millions; 9 is tens of thousands.

W – 2 Fifty-two billion, six hundred fifty million, seven thousand, eight hundred three

W – 3 3 400 008 241

W – 4
```
    565
 −113  once
 ─────
    452
 −113  twice
 ─────
    339
 −113  3 times
 ─────
    226
 −113  4 times
 ─────
    113
 −113  5 times
 ─────
      0
565 ÷ 113 = 5
```

W – 5 (a) 418   (b) 10^3 (c) 3375

(d) 12   (e) 9.5

W – 6 (a) 313 258   (b) 43 127

(c) 29 747 900   (d) 40 030

W – 7 (a) 41 922 Proof: 20 078 + 41 922 = 62 000

(b) 161 000 Proof: $\dfrac{161\ 000}{460} = 350$

(c) 301, R 155

Proof: 240 × 301 + 155 = 72 395.

FRACTIONS

F – 1 (a) Any fraction that has 10 as the upper part. Example: $\frac{10}{21}$ (b) $\frac{109}{8}$

F – 2 (a) $\frac{5}{7}$ (b) $\frac{49}{56}$ and $\frac{112}{128}$

F – 3 $44\frac{1}{6}$ F – 4 $7\frac{7}{16}$ F – 5 280

F – 6 (a) 25 (b) $\frac{1}{25}$ (c) $1\frac{31}{32}$

F – 7 45 is $\frac{3}{5}$ of 75 F – 8 128

DECIMALS

D – 1 (c) twenty-five and 340 ten-thousandths

D – 2 (a) 0.257 (b) 200.057

D – 3 $\frac{2}{125}$

D – 4 (a) 1 (b) 132.22 (c) 19.45 (d) 0.423

D – 5 (a) 0.006 (b) 0.005 52

D – 6 (a) 100 000 (b) 0.03 (c) 0.818

D – 7 (a) 1 place right (b) 4 places left

D – 8 (a) 1.0 (b) 186.45 (c) 1695

D – 9 (a) 15.24 (b) 21.143 (c) 0.093 75

D – 10 $3\frac{1}{3} \times \frac{3}{4} = 2\frac{1}{2}$

PERCENT

P – 1 (a) $\frac{13}{20}$ (b) $\frac{1}{20\ 000}$ (c) $\frac{5}{8}$ (d) $2\frac{3}{4}$

P – 2 (a) 4.63 (b) 0.0005 (c) 0.208 75

P – 3 (a) 31.25% (b) 403% (c) $8.\dot{3}\%$ or $8\frac{1}{3}\%$ (d) 7.5% or $7\frac{1}{2}\%$

P – 4 (a) 160 (b) 72 (c) 15.665 (d) 13.5

P – 5 (a) $45\frac{5}{11}\%$ (b) $33\frac{1}{3}\%$ (c) 0.1%

P – 6 (a) 187.5 (b) 350 (c) 25 000

P – 7 Inner and outer

P – 8 The value of t is 135.

P – 9 Rates: 25:95 = 15:57
Ratios: 25:15 = 95:57

ANSWERS FOR END-OF-CHAPTER TESTS -- PART ONE

WHOLE NUMBER TEST A

Addition.

(a) 454　　(b) 1325　　(c) 2449　　(d) 8684　　(e) 35

(f) 3925　　(g) 435　　(h) 202 211

Subtraction.

(a) 41　　(b) 335　　(c) 35 144　　(d) 4998　　(e) 16 109

(f) 101　　(g) 5309　　(h) 44 336

Multiplication.

(a) 3438　　(b) 109 620　　(c) 20 173 452　　(d) 709 504

(e) 1 986 580　　(f) 3 262 475　　(g) 425 918 010　　(h) 729 600

Division.

(a) 74　　(b) 21, R 22　　(c) 3, R 59　　(d) 300 002

(e) 1712, R 1　　(f) 1321, R 144　　(g) 999, R 31　　(h) 2280, R 199

WHOLE NUMBER TEST B

Addition.

(a) 414　　(b) 1162　　(c) 3038　　(d) 4864　　(e) 32

(f) 3391　　(g) 375　　(h) 121 771

Subtraction.

(a) 32　　(b) 366　　(c) 25 106　　(d) 998　　(e) 6009

(f) 301　　(g) 5309　　(h) 76 350

Multiplication.

(a) 3234　　(b) 191 835　　(c) 71 830 158　　(d) 467 062

(e) 1 487 008　　(f) 1 391 348　　(g) 403 574 214　　(h) 462 300

Division.

(a) 98, R 2　　(b) 21　　(c) 3, R 54　　(d) 400 002

(e) 1586, R 54　　(f) 1633, R 313　　(g) 999, R 21　　(h) 3189, R 40

WHOLE NUMBER TEST C

Addition.

(a) 411　　(b) 1189　　(c) 2096　　(d) 4556　　(e) 29

(f) 3394　　(g) 389　　(h) 139 796

Subtraction.

(a) 32　　(b) 467　　(c) 995　　(d) 44 127　　(e) 5813

(f) 311　　(g) 5029　　(h) 74 550

Multiplication.

(a) 5616　　(b) 209 761　　(c) 41 233 452　　(d) 475 931

(e) 230 016　　(f) 2 174 216　　(g) 402 753 654　　(h) 738 700

Division.

(a) 109　　(b) 22　　(c) 2, R 120　　(d) 300 005

(e) 1534, R 55　　(f) 1535, R 482　　(g) 999, R 64　　(h) 2282, R 176

FRACTION TEST A

(a) $2\frac{1}{4}$ (b) $\frac{4}{5}$ (c) $\frac{1}{2}$ (d) $1\frac{1}{2}$ (e) $3\frac{1}{6}$ (f) $10\frac{7}{12}$

(g) $15\frac{13}{24}$ (h) $\frac{2}{3}$ (i) $\frac{1}{6}$ (j) 10 (k) 32 (l) $\frac{3}{16}$

(m) $9\frac{3}{5}$ (n) $\frac{3}{4}$ (o) 81 (p) 36

FRACTION TEST B

(a) $2\frac{1}{2}$ (b) $\frac{7}{8}$ (c) $\frac{1}{5}$ (d) $1\frac{3}{8}$ (e) $15\frac{7}{10}$ (f) $13\frac{9}{10}$

(g) $14\frac{11}{24}$ (h) $\frac{3}{5}$ (i) $\frac{3}{10}$ (j) $22\frac{1}{2}$ (k) 18 (l) $\frac{5}{144}$

(m) $14\frac{2}{5}$ (n) 2 (o) 64 (p) 84

FRACTION TEST C

(a) $3\frac{1}{2}$ (b) $\frac{7}{12}$ (c) $\frac{1}{2}$ (d) $1\frac{2}{9}$ (e) $9\frac{7}{8}$ (f) $6\frac{5}{8}$

(g) $6\frac{19}{20}$ (h) $\frac{3}{5}$ (i) $\frac{1}{2}$ (j) $15\frac{3}{10}$ (k) $14\frac{2}{5}$ (l) $\frac{7}{32}$

(m) 12 (n) $1\frac{1}{2}$ (o) 49 (p) 64

DECIMAL TEST A

1. (a) 0.7   (b) 0.033   (c) 0.9375   2. 3.66   3. 0.41   4. $\frac{7}{20}$

5. 1   6. 0.008   7. 23.47   8. 3.71   9. 20.764   10. 34.05

11. 0.002 432   12. 0.36   13. 0.0005   14. 80

DECIMAL TEST B

1. (a) 0.3   (b) 0.061   (c) 0.5625   2. 2.24   3. 0.72   4. $\frac{16}{25}$

5. 1.1   6. 0.012   7. 35.603   8. 7.49   9. 29.777   10. 26.68

11. 0.000 108   12. 0.03   13. 0.0004   14. 70

DECIMAL TEST C

1. (a) 0.9   (b) 0.072   (c) 0.4375   2. 1.66   3. 0.61   4. $\frac{13}{20}$

5. 1.0 or 1   6. 0.009   7. 26.205   8. 7.31   9. 7.346   10. 29.84

11. 0.0003   12. 0.048   13. 0.0005   14. 700

PERCENT TEST A

1. (a) $\frac{1}{20}$ (b) $\frac{7}{200}$ (c) $\frac{3}{16}$ (d) $\frac{23}{500}$ 2. (a) 25%   (b) 350%   (c) 2.4%

3. (a) 240   (b) 39.4   (c) 4.8   4. (a) 40%   (b) $6\frac{1}{4}\%$

5. 63%   6. 20   7. 400   8. 2000

PERCENT TEST B

1. (a) $\frac{1}{20}$ (b) $\frac{3}{200}$ (c) $\frac{5}{16}$ (d) $\frac{4}{125}$ 2. (a) 75%   (b) 150%   (c) 1.6%

3. (a) 480   (b) 6.9   (c) 2.8   4. (a) 70%   (b) $1\frac{2}{3}\%$

5. 35%   6. 400   7. 20   8. 600

PERCENT TEST C

1. (a) $\frac{3}{20}$ (b) $\frac{7}{200}$ (c) $\frac{5}{16}$ (d) $\frac{9}{200}$ 2. (a) 25%   (b) 250%   (c) 2.4%

3. (a) 320   (b) 7.2   (c) 5.2   4. (a) 20%   (b) $8\frac{1}{3}\%$

5. 37%   6. 400   7. 50   8. 500

PRELIMINARY TEST – PART TWO

MEASUREMENT

PAGE	
125	M – 1 Complete the following tables. (a) 1 ft. = _12_ in. (b) 1 mi. = _____ rd. (c) 1 lb. = _16_ oz. (d) 1 gal. = _4_ qt.
127	M – 2 Change the following units of measure. (a) 3.5 ft. = _____ in. (b) 412 h = _3_ days (c) 75 sq. ft. = _____ sq. yd.
129	M – 3 (a) Subtract 4 lb. 11 oz. from 30 lb. 5 oz. (b) Divide 2 yd. 2 ft. 9 in. into three equal parts.
132	M – 4 Name the plane figures illustrated below.

133	M – 5 Match the formulas for area with the figures illustrated in M – 4. (a) $A = s^2$ (b) $A = lw$ (c) $A = \pi r^2$ (d) $A = \dfrac{bh}{2}$ (e) $A = h\dfrac{b_1 + b_2}{2}$
136	M – 6 Find the area of each of the following: (a) a rectangle that is 6 in. by 3 in. (b) a circle with diameter 8 in.
141	M – 7 (a) The angle at the base of an isosceles triangle is 48°. Write the sizes of the other two angles. (b) Find the length of the hypotenuse of a right triangle if the lengths of the other sides are 6 in. and 8 in.
144	M – 8 Find the circumference of a circle with diameter 10 cm.
147	M – 9 Find the volume of a rectangular box that is 2 ft. long, 16 in. wide, and 10 in. high.
151	M – 10 Complete the following: (a) 1 sq. yd. = _____ sq. ft. (b) 1 cu. ft. = _____ cu. in.
156	M – 11 Name the SI metric unit that measures each of the following: (a) distance (b) mass (c) liquid capacity
156	M – 12 Write the meaning of each of the following metric prefixes. (a) milli (b) centi (c) kilo

ALGEBRA

184   A--14 (a) Add $(4a + 7b) + (^-2a - 15b) + (a - 12b) + (^-20a + b)$

         (b) Subtract (i) ^-4x from 0   (ii) a from ^-3a

                    (iii) $2a + 6$ from $(^-4a - 8)$

185   A--15 Multiply.

         (a) $^-a \times ^-b$ (b) $^-a \times ^+a$

         (c) $4ab \times 0 \times 3ba$ (d) $a \times ^-1 \times ^-a$

187   A--16 Divide $\dfrac{6a^2bc}{4a}$ by $\dfrac{^-2ab}{12a^2}$

189   A--17 If $a = ^-2$, $b = 0.5$, and $c = ^-10$, find the value of

         (a) $3a^2 - bc$ (b) $\dfrac{5b}{c} \times c^3$

190   A--18 Solve for b in each of the following equations.

         (a) $b + 14 = 20$ (b) $0.2b + 0.4 = 0.6b$

         (c) $\dfrac{b}{4} = 21$ (d) $3b - 6 = 4b + 10$

         (e) $33b - (9b + 72) = 12b$ (f) $\dfrac{5}{8}b = 40$

         (g) $10 - 2.5b = 20 - 2b$

196   A--19 Rearrange the formula $A = \dfrac{bh}{2}$ so that the subject is h.

ANSWERS FOR PRELIMINARY TEST — PART TWO

MEASUREMENT

ALGEBRA

M – 1 (a) 12   (b) 320
       (c) 16   (d) 4

M – 2 (a) 42   (b) $17\frac{1}{6}$
       (c) $8\frac{1}{3}$

M – 3 (a) 25 lb. 10 oz.
       (b) 2 ft. 11 in.

M – 4 A rectangle   B square
       C parallelogram   D circle
       E triangle   F trapezoid

M – 5 (a) square   (b) rectangle
       (c) circle   (d) triangle
       (e) trapezoid

M – 6 (a) 18 sq. in.
       (b) 50.24 sq. in.

M – 7 (a) 48° and 84°   (b) 10 in.

M – 8 31.4 cm

M – 9 3840 cu. in. or $2\frac{2}{9}$ cu. ft.

M – 10 (a) 9   (b) 1728

M – 11 (a) metres etc.
       (b) grams etc.
       (c) litres etc.

M – 12 (a) one one-thousandth
       (b) one one-hundredth
       (c) 1000

M – 13 (a) 42 cm   (b) 3000 g
       (c) 5 L

M – 14 (a) cm²   (b) 546 728

M – 15 (a) 90 cm²   (b) 1000 cm³

A – 1 Like terms are (a) and (c).

A – 2 (a) ab (b) y^2

A – 3 (a) (i) 21   (ii) 21
         (iii) 21
      (b) (i) $^+4$ (ii) 0
         (iii) $^+2$ (iv) $^-4$

A – 4 $^-39$

A – 5 (a) $^+24$ (b) $^-56$
      (c) $^+40$

A – 6 (a) $^-25$ (b) $^+25$
      (c) 0   (d) $^+25$

A – 7 (a) 12   (b) $^-12$
      (c) $^+12$

A – 8 (a) $3n + 7$
      (b) $(400 + 10k + 25b)$

A – 9 (a) $8ab + 7b^2$ (b) $3a$

A – 10 $^-14a + 35b$

A – 11 $\dfrac{8a}{b^2c^3}$

A – 12 c^4

A – 13 (a) $a - b$ (b) $a + b$
      (c) ^-ab (d) $-\dfrac{a}{b}$

A – 14 (a) $^-17a - 19b$
      (b) (i) ^+4x (ii) ^-4a
         (iii) $^-6a - 14$

A – 15 (a) ab (b) $^-a^2$
      (c) 0   (d) a^2

A – 16 $^-9a^2c$

A – 17 (a) 17   (b) 250

A – 18 (a) $b = 6$ (b) $b = 1$
      (c) $b = 84$ (d) $b = ^-16$
      (e) $b = 6$ (f) $b = 64$
      (g) $b = ^-20$

A – 19 $h = \dfrac{2A}{b}$

CHAPTER FIVE:**MEASUREMENT**

*Most countries now use the metric system of measurement.
Canada and the United States are in the process of converting
from the British (or Imperial) system to the metric, so both
systems are in use at the present time.*

*This chapter deals with British measures first, followed by a
section on SI metric.*

ITEMS IN CHAPTER
PART ONE

PART TWO

ITEM M – 1 SOME TABLES OF MEASURE:
BRITISH AND METRIC

This is just a partial list of the most common measures.

LINEAR MEASUREMENT (DISTANCE OR LENGTH)

British units: smaller to larger			Metric units: smaller to larger		
Singular	*Plural*	*Abbreviation*	*Singular*	*Plural*	*Symbol*
inch	inches	in.	millimetre	millimetres	mm
foot	feet	ft.	centimetre	centimetres	cm
yard	yards	yd.	decimetre	decimetres	dm
rod	rods	rd.	metre	metres	m
mile	miles	mi.	decametre	decametres	dam
			hectometre	hectometres	hm
			kilometre	kilometres	km

1 ft. = 12 in.	10 mm = 1 cm	1 mm = 0.001 m
1 yd. = 3 ft. or 36 in.	10 cm = 1 dm	1 cm = 0.01 m
1 rd. = $5\frac{1}{2}$ yd. or $16\frac{1}{2}$ ft.	10 dm = 1 m	1 dm = 0.1 m
1 mi. = 320 rd. or 1760 yd.	10 m = 1 dam	1 dam = 10 m
or 5280 ft.	10 dam = 1 hm	1 hm = 100 m
1 acre = 160 sq. rd.	10 hm = 1 km	1 km = 1000 m

MEASUREMENT OF MASS OR WEIGHT

British units: smaller to larger			Metric units: smaller to larger		
Singular	*Plural*	*Abbreviation*	*Singular*	*Plural*	*Symbol*
ounce	ounces	oz.	milligram	milligrams	mg
pound	pounds	lb.	centigram	centigrams	cg
hundred-	hundred-	cwt.	decigram	decigrams	dg
weight	weights		gram	grams	g
ton	tons	T.	decagram	decagrams	dag
			hectogram	hectograms	hg
			kilogram	kilograms	kg

1 lb. = 16 oz.	10 mg = 1 cg	1 mg = 0.001 g
1 cwt. = 100 lb.	10 cg = 1 dg	1 cg = 0.01 g
1 T. = 20 cwt. or 2000 lb.	10 dg = 1 g	1 dg = 0.1 g
	10 g = 1 dag	1 dag = 10 g
1 lb. = 454 g.	10 dag = 1 hg	1 hg = 100 g
	10 hg = 1 kg	1 kg = 1000 g

MEASUREMENT OF LIQUID CAPACITY

British units: smaller to larger			Metric units: smaller to larger		
Singular	*Plural*	*Abbreviation*	*Singular*	*Plural*	*Symbol*
cup	cups	c.	millilitre	millilitres	mL
pint	pints	pt.	centilitre	centilitres	cL
quart	quarts	qt.	decilitre	decilitres	dL
gallon	gallons	gal.	litre	litres	L
			decalitre	decalitres	daL
			hectolitre	hectolitres	hL
			kilolitre	kilolitres	kL

1 pt. = 2 c. 1 qt. = 2 pt. or 4 c. 1 gal. = 4 qt. or 8 pt.	10 mL = 1 cL 10 cL = 1 dL 10 dL = 1 L 10 L = 1 daL 10 daL = 1 hL 10 hL = 1 kL	1 mL = 0.001 L 1 cL = 0.01 L 1 dL = 0.1 L 1 daL = 10 L 1 hL = 100 L 1 kL = 1000 L

ITEM M – 2 CONVERSION OF UNITS
WITHIN THE BRITISH SYSTEM

STEPS

One: *Identify the conversion factor.*

Two: *Multiply or divide by the conversion factor.*

EXPLANATION

Step 1: The conversion factor is the number that relates two units of measure; it tells how many of the smaller unit are required to make one of the larger unit.

For example,

The conversion factor between feet and inches is 12	$\left(12\,\text{in.} = 1\,\text{ft.}\right)$
The conversion factor between ounces and pounds is 16	$\left(16\,\text{oz.} = 1\,\text{lb.}\right)$
The conversion factor that relates pints and gallons is 8	$\left(8\,\text{pt.} = 1\,\text{gal.}\right)$

Step 2: In deciding whether to multiply or divide, follow this rule:

Multiply to convert *larger* units to *smaller* units.

Divide to convert *smaller* units to *larger* units.

For example,

To convert feet to inches is to convert *larger* to *smaller* →*Multiply*

To convert seconds to minutes is to convert *smaller* to *larger* → *Divide*

Example A

How many feet are there in 275 inches?

STEPS

One: The conversion factor is 12.

Two: Smaller (inches) to larger (feet): Divide

There are $22\frac{11}{12}$ ft. in 175 in.

$$
\begin{array}{r}
22 \\
12\overline{)275} \\
24 \\
\hline
35 \\
24 \\
\hline
11
\end{array}
$$

Example B

Change 3.5 sq. yd. to square feet.

STEPS

One: The conversion factor is 9.

Two: Larger (square yards) to smaller (square feet): Multiply

3.5 sq. yd. = 31.5 sq. ft.

$$
\begin{array}{r}
3.5 \\
9 \\
\hline
31.5
\end{array}
$$

EXERCISE M – 2

1. Name the conversion factor that relates each pair. Use the tables on page 125 if necessary.

   (a) feet and yards
   (b) yards and feet
   (c) square rods and acres
   (d) quarts and gallons
   (e) days and hours
   (f) cups and gallons
   (g) pounds and ounces
   (h) miles and feet
   (i) hours and seconds
   (j) tons and pounds
   (k) weeks and years
   (l) bushels and pecks
   (m) square inches and square feet
   (n) cubic feet and cubic yards

2. In each case, tell whether the change is from smaller to larger or vice versa, whether you would multiply or divide, and by what number.

   (a) To change miles to yards.
   (b) To change days to hours.
   (c) To change seconds to minutes.
   (d) To change sq. ft. to sq. yd.
   (e) To change quarts to gallons.
   (f) To change cu. ft. to cu. in.

3. (a) How many pints are there in 9.25 gal.?
   (b) Find the number of tons in 33 175 lb.
   (c) Change 396 sq. in. to square feet.
   (d) How many pints in 65 cups?
   (e) Change $7\frac{5}{12}$ ft. to inches.
   (f) Find the number of miles in 3488 rd.

4. Convert each of the following to yards.

   (a) 30 in.   (b) 17.5 ft.   (c) 88 rd.   (d) 4428 in.

ANSWERS FOR EXERCISE M – 2

1. (a) 3   (b) 3   (c) 160   (d) 4   (e) 24   (f) 16   (g) 16
   (h) 5280   (i) 3600   (j) 2000   (k) 52   (l) 4   (m) 144   (n) 27
2. (a) Larger to smaller; multiply by 1760.   (b) Larger to smaller; multiply by 24.
   (c) Smaller to larger; divide by 60.   (d) Smaller to larger; divide by 9.
   (e) Smaller to larger; divide by 4.   (f) Larger to smaller; multiply by 1728.
3. (a) 74   (b) 16.5875 or $16\frac{47}{80}$ (c) 2.75 or $2\frac{3}{4}$ sq. ft.
   (d) $32\frac{1}{2}$ (e) 89 in.   (f) 10.9 or $10\frac{9}{10}$
4. (a) $\frac{5}{6}$ yd.   (b) $5\frac{5}{6}$ yd.   (c) 484 yd.   (d) 123 yd.

ITEM M – 3 OPERATIONS WITH DENOMINATE NUMBERS

A DENOMINATE NUMBER is one that refers to a specific unit of measurement. Examples are: 32 ft., 7 h 45 min, 112 m², etc.

These numbers may be added, subtracted, multiplied or divided. There are two approaches, both illustrated below.

Method 1:

This method requires that you deal with each unit of measurement separately, and then simplify the final answer.

EXAMPLE OF ADDITION

First copy the numbers in columns.	3 yd. 2 ft. 7 in.
	4 yd. 1 ft. 11 in.
	5 yd. 2 ft. 9 in.
Second, add each column separately.	12 yd. 5 ft. ~~27~~ in.
Change 27 in. to 2 ft. 3 in.	+2 ft. 3 in.
Combine 2 ft. 3 in. with previous answer.	12 yd. ~~7~~ ft. 3 in.
Change 7 ft. to 2 yd. 1 ft.	+2 yd. 1 ft.
Combine this with other answer.	14 yd. 1 ft. 3 in.

Method 2:

This method involves changing the units of measurement to a mixed number with the largest unit as the whole number, and the other(s) as a fraction of the largest unit.

SAME EXAMPLE OF ADDITION

3 yd. 2 ft. 7 in. = 3 yd. + (24 + 7) in. = 3 yd. + 31 in., or $3\frac{31}{36}$ yd.

4 yd. 1 ft. 11 in. = 4 yd. + (12 + 11) in. = 4 yd. + 23 in., or $4\frac{23}{36}$ yd.

5 yd. 2 ft. 9 in. = 5 yd. + (24 + 9) in. = 5 yd. + 33 in., or $5\frac{33}{36}$ yd.

Add: $3\frac{31}{36} + 4\frac{23}{36} + 5\frac{33}{36} = 12\frac{87}{36} = 14\frac{15}{36}$

Convert answer to original form.

$14\frac{15}{36}$ yd. = 14 yd. 15 in., or 14 yd. 1 ft. 3 in. (final answer)

EXAMPLE OF SUBTRACTION

Subtract 2 lb. 12 oz. from 7 lb. 3 oz.	
Method 1	**Method 2**
7 lb. 3 oz. = 6 lb. 19 oz. 2 lb. 12 oz. = 2 lb. 12 oz. Answer: 4 lb. 7 oz.	$7\frac{3}{16} - 2\frac{12}{16} = 6\frac{19}{16} - 2\frac{12}{16} = 4\frac{7}{16}$ $4\frac{7}{16}$ lb. = 4 lb. 7 oz.

EXAMPLE OF MULTIPLICATION

Multiply 3 sq. yd. 7 sq. ft. by 11.	
Method 1	**Method 2**
3 sq. yd. 7 sq. ft. $\times$ 11 33 sq. yd. 77 sq. ft. 8 sq. yd. 5 sq. ft. 41 sq. yd. 5 sq. ft.	$3\frac{7}{9} \times 11 = \frac{34}{9} \times \frac{11}{1} = \frac{374}{9} = 41\frac{5}{9}$ $41\frac{5}{9}$ sq. yd. = 41 sq. yd. 5 sq. ft.

EXAMPLE OF DIVISION

Divide 7 gal. 3 qt. 1 pt. by 3	
Method 1	**Method 2**
<div>　　2 gal.　　2 qt.　　1 pt. 3) 7 gal.　　3 qt.　　1 pt. 　 6 gal. 　 1 gal. = 4 qt. 　　　　　 7 qt. 　　　　　 6 qt. 　　　　　 1 qt. = 2 pt. 　　　　　　　　 3 pt. 　　　　　　　　 3 pt. 　　　　　　　　　 0</div>	7 gal. 3 qt. 1 pt. = 7 gal. + 7 pt., or $7\frac{7}{8}$ gal. $7\frac{7}{8} \div 3 = \frac{\overset{21}{\cancel{66}}}{8} \times \frac{1}{\underset{1}{\cancel{3}}} = \frac{21}{8} = 2\frac{5}{8}$ $2\frac{5}{8}$ gal. = 2 gal. 5 pt. 　　　　　 = 2 gal. 2 qt. 1 pt.

EXERCISE M - 3

Solve the following using the method that you prefer.

1. Add: 3 h 40 min 37 s + 7 h 50 min + 6 h 38 min 59 s
2. Subtract 1 gal. 3 qt. from 6 gal. 2 qt.
3. Multiply 20 mi. 1500 yd. by 4.
4. Divide 20 d 15 h by 3.
5. Add 4 T. 1600 lb., 12 T. 300 lb., 1800 lb., and 8 T. 950 lb.
6. Subtract 5 h 42 min 50 s from 11 h 3 min 21 s.
7. Multiply 21 rd. 7 yd. by 6.
8. Divide 15 yd. 2 ft. 9 in. by 3.

ANSWERS FOR EXERCISE M - 3

1. 18 h 9 min 36 s

2. $6\frac{2}{4} - 1\frac{3}{4} = 5\frac{6}{4} - 1\frac{3}{4} = 4\frac{3}{4} = 4$ gal. 3 qt.

3. 83 mi. 720 yd.

4. 6 d 21 h

5.
```
     4 T. 1600 lb.
    12 T.  300 lb.
           1800 lb
  + 8 T.  950 lb.
  _____
    24 T. 4650 lb.
    26 T.  650 lb.
```

6. 11 h 3 min 21 s = 10 h 62 min 81 s
$$\underline{\ 5\ h\ 42\ min\ 50\ s}$$
$$5\ h\ 20\ min\ 31\ s$$

7.
```
  21 rd. 7 yd.
       × 6
  _____
 126 rd. 42 yd.  = 133 rd. 3½ yd.
```

8. 5 yd. 11 in.

ITEM M – 4   PLANE FIGURES: DEFINITIONS AND PERIMETER

SOME DEFINITIONS:

1. **Perimeter** is the length of a closed curve (e.g., the distance around a circle) or the distance around a polygon.
2. **Circle** is a closed curve in which all points are equidistant from the centre.
3. **Polygon** is a closed plane geometric figure having three or more sides. Some examples of polygons are:
   (a) Triangle — A three-sided polygon
   (b) Quadrilateral — A four-sided polygon
   (c) Pentagon — A five-sided polygon
   (d) Hexagon — A six-sided polygon

The following section shows some plane geometric figures and how to calculate their perimeters.

Name and Diagram	Definition	Dimensions	Formula for Perimeter
Rectangle	A quadrilateral with all angles right angles	length width	$P = 2(l + w)$
Square	A rectangle with all sides equal.	side	$P = 4s$
Parallelogram	A quadrilateral with opposite sides parallel	-----	No formula. Add the 4 sides.
Rhombus	A parallelogram with all sides equal, and with oblique angles	side	$P = 4s$
Trapezoid	A quadrilateral with one pair of sides parallel	side	No formula. Add the lengths of the 4 sides.
Circle	A closed curve in which all points are equidistant from the centre	Radius or diameter	$C = 2\pi r$ or $C = \pi d$
Triangle	A three-sided polygon	side	No formula. Add the lengths of three sides.

See Exercise M – 4 on page 134.

ITEM M – 5   AREA: MEANING AND FORMULAS

The area of a plane figure is the number of square units enclosed within its boundaries. Here are some diagrams of plane figures showing the areas that they represent. Count (and estimate) the number of squares in each diagram.

SQUARE

Side is $3\frac{1}{2}$ ft.

Area is $12\frac{1}{4}$ sq. ft.

RECTANGLE

Length is 7 ft.
Width is 3 ft.
Area is 21 sq. ft.

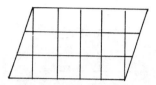

PARALLELOGRAM

Base is 5 ft.
Height is 3 ft.
Area is 15 sq. ft.

TRIANGLE

Base is 3 ft.
Height is 4.5 ft.

Area is 6.75 sq. ft.

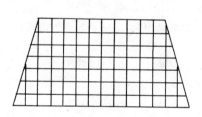

TRAPEZOID

Base 1 is 14 in.
Base 2 is 10 in.
Height is 7 in.
Area is 84 sq. in.

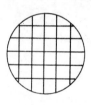

CIRCLE

Radius is 3 ft.

Area is 28.26 sq. ft.

FORMULAS FOR AREAS OF PLANE FIGURES

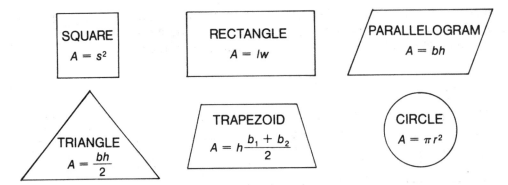

The area of a *square* is side squared.

The area of a *rectangle* is length times width.

The area of a *parallelogram* is base times height.

The area of a *triangle* is one-half of the base times the height.

The area of a *trapezoid* is half the sum of the bases times the height.

The area of a *circle* is 3.14 times the radius times the radius.

EXERCISE M -- 4 and 5

1. From the list below, select the name of the plane figure that matches the descriptive phrase.

Triangle   Square   Parallelogram   Rectangle

Rhombus   Circle   Trapezoid

(a) To find its area, you multiply length by width.

(b) It is a quadrilateral with one pair of parallel sides.

(c) The formula for its perimeter is $P = 4s$.

(d) It has two bases of different lengths.

(e) It has half the area of a parallelogram with the same dimensions.

(f) The formula for its area is $A = \pi r^2$.

(g) All sides are equal and all angles are right angles.

(h) All sides are equal but angles are not right angles.

(i) The formula for its area is

    (i) $A = lw$ (ii) $A = bh$ (iii) $A = \dfrac{bh}{2}$ (iv) $A = s^2$

(j) It is a closed curve.

2. Identify each of the following statements as either true or false. Be able to support your answer.

   (a) A polygon is a quadrilateral.

   (b) A quadrilateral is a polygon.

   (c) A rectangle is a parallelogram.

   (d) A parallelogram is a rectangle.

   (e) A parallelogram has 2 right angles and 2 acute angles.

   (f) πr^2 is the same as πd.

   (g) If you know the circumference of a circle, you can find its diameter.

   (h) Measurement of length is called linear measure.

   (i) 1 sq. ft. $= 144$ in.

   (j) The circumference of a circle is roughly $6\frac{1}{3}$ times the radius.

ANSWERS FOR EXERCISE M -- 4 and 5

1. (a) rectangle   (b) trapezoid   (c) square or rhombus   (d) trapezoid
   (e) triangle   (f) circle   (g) square   (h) rhombus
   (i) (*i*) rectangle   (*ii*) parallelogram   (*iii*) triangle   (*iv*) square
   (j) circle

2. (a) False. A polygon can be 3 sided, 4 sided, 5 sided etc. but a quadrilateral is only 4 sided.
   (b) True.
   (c) True. A rectangle fits the definition of a parallelogram.
   (d) False. Only true if the parallelogram happens to have square corners.
   (e) False. This is an impossible combination.
   (f) False. $2\pi r = \pi d$
   (g) True. Divide by π.   (h) True.
   (i) False. 1 sq. ft. $= 144$ sq. in.   (j) True; circumference is 3.14 times diameter or 6.28 times radius.

ITEM M – 6   FINDING THE AREA OF PLANE FIGURES

STEPS

One: *Identify the shape and the corresponding formula.*

Two: *Check that all dimensions are in the same linear unit, and then apply the formula.*

Example A

Find the area of a board fence which is 10 yd. long and 4 ft. high.

STEPS

One: The fence is rectangular in shape.

The formula is $A = lw$ (in this case lh)

Two: $l = 30$ (change 10 yd. to 30 ft.) $h = 4$
$lh = 30 \times 4 = 120$

The area of the fence is 120 sq. ft.

MORE EXAMPLES

1. Find the area of a square with side 2 inches.

   $A = s^2; s = 2,$ $s^2 = 4$

   Area is 4 sq. in.

2. Find the area of a parallelogram with base 4.5 m and height 2.2 m.

   $A = bh; b = 4.5$ and $h = 2.2,$ $4.5 \times 2.2 = 9.9$

   Area is $9.9\,m^2$ (Note: In metric terminology, m^2 is preferable to sq. m)

3. Find the area of a triangle with base 12 in. and height $7\frac{1}{4}$ in.

   $A = \dfrac{bh}{2};$ $b = 12$ and $h = 7\frac{1}{4},$ $\dfrac{12 \times 7\frac{1}{4}}{2} = 43\frac{1}{2}.$

   Area is $43\frac{1}{2}$ sq. in.

4. Find the area of a trapezoid with the following dimensions:
   Height is 9''; bases are 16'' and 10''.

   $A = h\dfrac{b_1 + b_2}{2};$ $h = 9,$ $b_1 = 16,$ and $b_2 = 12$

   $9\left(\dfrac{16 + 12}{2}\right) = 126$

   Area is 126 sq. in.

5. Find the area of a circle with diameter 20 in.

   $A = \pi r^2;$ $\pi = 3.14$ and $r = 10,$ (if diameter is 20, radius is 10)

   $3.14 \times 10^2 = 314$

   Area is 314 sq. in.

Sometimes there is more than one shape involved in finding the area of a given figure. Consider the following example:

Find the area of the figure illustrated below.

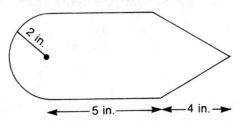

The figure is composed of

(a) a rectangle   +   (b) a triangle   +   (c) half of a circle

$A = lw$ + $\dfrac{bh}{2}$ + $\frac{1}{2}$ of πr^2

$A = 4 \times 5$ + $\dfrac{4 \times 4}{2}$ + $\frac{1}{2}$ of 3.14×2^2

$A = 20$ +   8   +   6.28

The area of the figure is 34.28 sq. in.

... or this example:

Find the area of the shaded portion.
The area of the shaded part is:

this area [ ] − this area △

lw (12×8) $-\dfrac{bh}{2}$ $\left(\dfrac{10 \times 8}{2}\right)$

96   −   40

The area of the shaded portion is 56 sq. in.

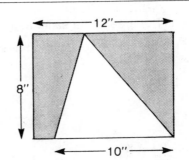
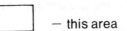

EXERCISE M -- 5 and 6

Finding perimeter and area.

1. The formula for the area of a rectangle is $A = lw$.

   If any two of these elements are known, the third can be calculated.

   If $A = lw$, $w = \dfrac{A}{l}$ and $l = \dfrac{A}{w}$.   Use this information to complete the following table.

	Length	Width	Perimeter	Area
(a)	10 in.	2.5 in.		
(b)	14 cm			42 cm²
(c)	1.5 m		5 m	
(d)		0.3 m		0.15 m²
(e)	twice the width			32 sq. in.

2. Using whole numbers only, give three sets of dimensions of rectangles that would each have an area of 24 sq. in.

3. Find the length of one side of each of the squares described below.

   (a) The area is 4 cm².   (b) The area is 25 sq. ft.

   (c) The perimeter is 20.12 m.

   (d) The perimeter is the same as the circumference of a circle that has a diameter of $3\frac{1}{2}$ in.   (Use $\pi = \frac{22}{7}$.)

   (e) The area is the same as that of a rectangle that is 8 ft. by 18 ft.

4. Find the number of square units represented by the shaded part of each diagram below.

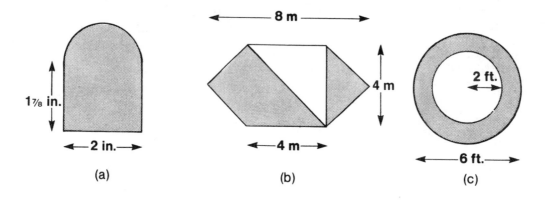

(a)   (b)   (c)

5. Using the formula for the area of a trapezoid $A = h\left(\dfrac{b_1 + b_2}{2}\right)$, complete the table below.

	Base$_1$	Base$_2$	Height	Area
(a)	10 cm	14 cm	6 cm	
(b)	$\frac{1}{2}$ ft.	8 in.	10 in.	in sq. in.
(c)	1.5 m	2 m	70 cm	
(d)	12 in.	8 in.		90 sq. in.
(e)	0.6 m	0.4 m		0.5 m²

6. Complete the following chart.

	Plane figure	Dimensions	Perimeter	Area
(a)	Square	side = 1.3 m		
(b)	Parallelogram	base = $3\frac{1}{2}$ in., height = 2 in.	——	
(c)	Triangle	base = 10.4 cm, height = 16 cm	——	
(d)	Rectangular floor	6 yd. by 15 ft.		
(e)	Piece of cloth	1 yd. by 30 in.		
(f)	Circle	radius 5 in.		
(g)	Circle	diameter 8 ft.		
(h)	Circle	radius 1 ft. 3 in.		

ANSWERS FOR EXERCISE M - 5 and 6

1. (a) Perimeter = 2(10 + 2.5) in., or 25 in. Area = 10 × 2.5, or 25 sq. in.

   (b) Width = 42 ÷ 14, or 3 cm Perimeter = 2(14 + 3), or 34 cm

   (c) Width = half of (5 − 3), or 1 m Area = 1.5 × 1 or 1.5 m²

   (d) Length = 0.15 ÷ 0.3 or 0.5 m Perimeter = 2(0.5 + 0.3), or 1.6 m

   (e) Two numbers, one twice as great as the other, have a product of 32. The numbers must be 8 and 4. The length is 8 in.; the width is 4 in. The perimeter is 2(8 + 4) = 24 in.

2. Possible answers: 24 in. by 1 in.; 12 in. by 2 in.; 8 in. by 3 in.; or 6 in. by 4 in.
3. Since the area of a square is side squared, a side must be the square root of the area.
   (a) $\sqrt{4} = 2$; side is 2 cm   (b) $\sqrt{25} = 5$; side is 5 ft.
   (c) Since the perimeter is $4s$, a side must be (perimeter $\div$ 4) $20.12 \div 4 = 5.03$; side is 5.03 m.
   (d) Circumference $= \pi d$, or $\frac{22}{7} \times 3\frac{1}{2}$.   Circumference $= 11$ in.
       Perimeter also equals 11 in.   Therefore one side is $11 \div 4$, or $2\frac{3}{4}$ in.
   (e) Area of rectangle is 144 sq. ft. Area of square is 144 sq. ft.
       Side is $\sqrt{144}$, or 12 ft.
4. (a) The figure is a rectangle $+$ half of a circle.
       Area of the rectangle is $2 \times 1\frac{7}{8}$, or 3.75 sq. in.
       Area of half-circle is $\dfrac{3.14 \times 1 \times 1}{2}$, or 1.57 sq. in.
       Total area of figure is 5.32 sq. in.

   (b) There are many ways to do this; one is to regard the total figure as a four metre square with two triangles that together form another square half its size. From this total area, subtract the unshaded part which is a triangle with base 4 m and height 4 m.

       Area of shaded part is $4^2 + \dfrac{4^2}{2} - \dfrac{4 \times 4}{2} = 16 + \dfrac{16}{2} - 8$, or 16 m².

   (c) The shaded part is the area of the larger circle minus the area of the smaller circle. Since the diameter of the larger circle is 6 ft., its radius must be 3 ft.
       Area of the shaded part is $3.14 \times 3^2 - 3.14 \times 2^2$, or 15.7 sq. ft.
       The short cut here is $3.14(9 - 4)$, or 3.14×5.

5. (a) $\dfrac{10 + 14}{2} \times 6 = 72$ cm²   (b) $\dfrac{6 + 8}{2} \times 10 = 70$ sq. in.

   (c) $\dfrac{1.5 + 2}{2} \times 0.7 = 1.225$ m²   (d) $\dfrac{12 + 8}{2} = 10$; $10 \times$ height $= 90$

       Height must be 9 in.

   (e) $\dfrac{0.6 + 0.4}{2} = 0.5$; $0.5 \times ? = 0.5$. Height is 1 m.

6. Perimeters: (a) 5.2 m; (d) 66 ft. or 22 yd.; (e) 132 in.; (f) 31.4 in.;
                 (g) 25.12 ft.; (h) 7.85 ft.

   Areas: (a) 1.69 m²; (b) 7 sq. in.; (c) 83.2 cm²; (d) 30 sq. yd. or 270 sq. ft.;
                (e) 1080 sq. in.; (f) 78.5 sq. in.; (g) 50.24 sq. ft.;
                (h) 4.906 sq. ft. or 706.5 sq. in.

For additional practice with area see page 221.

ITEM M – 7 THE TRIANGLE

A triangle is a three-sided polygon.
Triangles may be classified according to sides:

Equilateral Triangle
(all sides are equal)

Isosceles Triangle
(two sides are equal)

Scalene Triangle
(there are no equal sides)

Types of angles:

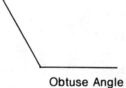

Right Angle
(exactly 90°)

Obtuse Angle
(more than 90°)

Acute Angle
(less than 90°)

Triangles may be classified according to angles:

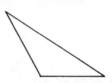

Right Triangle
(1 right angle)

Obtuse Triangle
(1 obtuse angle)

Acute Triangle
(3 acute angles)

Equiangular Triangle
(3 equal angles)

The *base* of a triangle may be any of the three sides but is usually the side on which the triangle rests. The *base* of an isosceles triangle is the unequal side.

The *vertex* of a triangle is the angle that is opposite the base.

The *altitude* of a triangle (or the height) is the perpendicular distance from the base to the vertex.

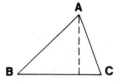 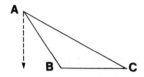 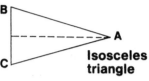

In each case, *A* is the vertex; *BC* is the base; and the broken line is the altitude.

- The *hypotenuse* of a right triangle is the side opposite the right angle.
- The *sum of the angles* of a triangle is 180°.
- The *perimeter* of a triangle is the sum of the lengths of the three sides.
- The *area* of a triangle is half the product of the base and the height $A = \dfrac{b \times h}{2}$

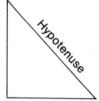

- *Similar triangles* have the same shape;
  i.e. they have equal angles, and sides that are respectively proportional.

THE RIGHT TRIANGLE

A right triangle is one that has a right angle (an angle of 90°). The side opposite the right angle is the hypotenuse.

There is a constant relationship between the hypotenuse of a right triangle and the other two sides. This relationship is stated in the Pythagorean Theorem, which says:

> *The square of the hypotenuse of a right triangle is equal to the*
> *sum of the squares of the other two sides.*

Using this theorem, it is possible to find the length of any side of a right triangle if the lengths of two sides are known.

Example A

Find the length of the hypotenuse.

STEPS

First: Square the other two sides. $3^2 = 9; 4^2 = 16$

Second: Find their sum. $9 + 16 = 25$

Third: Find the square root of the sum. $\sqrt{25} = 5$

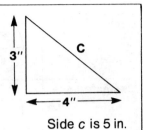

Side c is 5 in.

Example B

Find the length of a side (not the hypotenuse).

STEPS

First: Square the hypotenuse and square the known side. $15^2 = 225; 12^2 = 144$

Second: Find the difference. $225 - 144 = 81$

Third: Find the square root of the difference. $\sqrt{81} = 9$

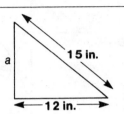

Side a is 9 in.

In the two examples above, it was possible to find the square roots using your knowledge of tables. For the next example, two methods of finding the square root are shown. You should not be unduly concerned if you are unable to find square root at the present time, because it will be re-taught in any class that requires it.

Example C

In a right triangle, the hypotenuse is 35 in. and one of the sides is 21 in. Find the length of the other side.

STEPS

One: $35^2 = 1225; 21^2 = 441$

Two: $1225 - 441 = 784$

Three: $\sqrt{784} = 28$

The length of the other side is 28 in.

FINDING THE SQUARE ROOT OF 784

Method 1	Method 2
$\begin{array}{r}2\quad\ \ 8\\ \overline{\sqrt{7\ 8\ 4}}\\ 4\ \ \ \ \ \ \ \\ \hline 4\ 8\ \ \ 3\ 8\ 4\\ 3\ 8\ 4\\ \hline 0\end{array}$	$\begin{array}{r}2)\ 7\ 8\ 4\\ 2)\ 3\ 9\ 2\\ 2)\ 1\ 9\ 6\\ 2)\quad 9\ 8\\ 7)\quad\ \ 4\ 9\\ 7)\quad\quad\ 7\\ 1\end{array}$
	$2 \times 2 \times 7 = 28$

ITEM M – 8 THE CIRCLE

Circle		A circle is a closed curve in which all points are equidistant from the centre.
Circumference		The circumference is the length of the circle, or the distance around the enclosed space.
Radius		The radius is a straight line from the centre of the circle to any point on the circumference.
Diameter		The diameter is a straight line passing through the centre of the circle and touching the circumference at two points.
Chord		A chord is a straight line connecting any two points on the circumference.
Semicircle		A semicircle is half of a circumference.
Concentric Circles		Concentric circles have a common centre.

π (pronounced "pi") is the Greek letter which represents the relationship between the circumference of a circle and its diameter. In any circle, regardless of size, the circumference is roughly $3\frac{1}{7}$ times as long as the diameter. Therefore π equals $\frac{c}{d}$, or circumference divided by diameter.

When calculating with π, you may use the rough equivalent $3\frac{1}{7}$ (or $\frac{22}{7}$), but a closer approximation is 3.14. If you want greater accuracy still, use 3.1416. For calculations requiring a very high degree of accuracy, π has been worked out to hundreds, even thousands, of decimal places.

Formula for circumference: $C = \pi d$ (Circumference = 3.14 × diameter)

Formula for area: $A = \pi r^2$ (Area = 3.14 × radius × radius)

EXERCISE M – 7

1. Identify the following triangles as equilateral, scalene, or isosceles.
   (a) Sides are 3 cm, 4 cm, and 5 cm.   (b) Each angle is 60°.
   (c) 2 sides are equal.
2. For each triangle, $\angle A$ and $\angle B$ are given. Find $\angle C$.
   (a) $\angle A = 47°$ $\angle B = 22°$ (b) $\angle A = 69°$ $\angle B = 94°$
   (c) $\angle A = 110°$ $\angle B = 38°$
3. Find the perimeter of each triangle:
   (a) Sides are 6 in., 8 in. and 9.5 in.
   (b) An equiangular triangle with one side 10 cm.
4. Find the area of each triangle:
   (a) Base is 4.5 m and height is 9.3 m.
   (b) Base is half the height of $4\frac{1}{4}$ in.
5. Find the perimeter and area of each right triangle illustrated. (First you use the Pythagorean Theorem to find the missing length.)

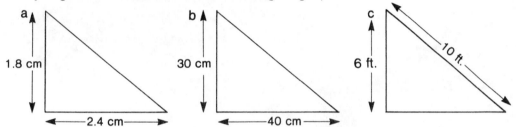

ANSWERS FOR EXERCISE M – 7

1. (a) scalene   (b) equilateral   (c) isosceles
2. (a) 111°   (b) 17°   (c) 32°
3. (a) 23.5 in.   (b) 30 cm   4. (a) 20.925 m²   (b) 4.516 sq. in.
5. (a) Side is $\sqrt{1.8^2 + 2.4^2} = \sqrt{9} = 3$ cm.
     Perimeter is $1.8 + 2.4 + 3$ or 7.2 cm.
     Area is $\dfrac{1.8 \times 2.4}{2} = 2.16$ cm²
   (b) Side is $\sqrt{30^2 + 40^2}$, or $\sqrt{2500}$ or 50 cm.
     Perimeter is 120 cm.
     Area is 600 cm².
   (c) Side is $\sqrt{10^2 - 6^2}$, or $\sqrt{64}$, or 8 ft.
     Perimeter is 24 ft.
     Area is 24 sq. ft.

EXERCISE M -- 8

1. Name the part of the circle that is defined.

    (a) A straight line passing through the centre and touching the circumference at two points.

    (b) The perimeter or the distance around a circle.

    (c) A straight line from the centre to the circumference.

2. What is the radius of each of the following circles?

    (a) The diameter is 27 cm.

    (b) The circumference is 314 in.

3. Find the diameter when:

    (a) the radius is 2.5 ft.   (b) the circumference is 157 cm.

4. Using the formula $A = \pi r^2$, find the area of each circle in questions 2 and 3 above.

5. Find the difference between two circles with diameters of 7 in. and 8 in., respectively, in (a) circumference   (b) area.

ANSWERS FOR EXERCISE M – 8

1. (a) diameter   (b) circumference   (c) radius

2. (a) 13.5 cm   (b) $\dfrac{314}{2 \times 3.14} = 50$ in.

3. (a) 5 ft.   (b) 50 cm

4. 572.265 cm²   7850 sq. in.   19.625 sq. ft.   1962.5 cm²

5. (a) Circumference: $(25.12 - 21.98) = 3.14$ in.   (b) Area: difference is 11.775 sq. in.
    or $3.14(8 - 7)$.

ITEM M – 9 VOLUME OR CAPACITY

> The volume of an object is the amount of space that it occupies.

Volume is measured in cubic units: cubic inches, cubic feet, cubic centimetres, etc.

When we say that the volume of an object is 120 cm^3, we mean that the object occupies the same amount of space that would be taken up by 120 cubes, each measuring 1 cm by 1 cm by 1 cm. Think of 120 sugar cubes (which are roughly the size of a cubic centimetre) packed in rows and layers with no space between.

STEPS TO FIND THE VOLUME OF A SOLID

One: *Identify the shape and the formula.*

Two: *Check that all dimensions are in the same unit.*

Three: Apply the formula.

EXAMPLES OF SOME SOLIDS

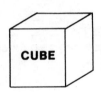

Cube
All dimensions are equal; all faces are squares.
Dimension: side (*s*)
Formula for volume: $V = s^3$
Volume is side cubed.

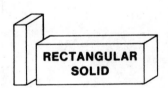

Rectangular Solid
Dimensions: length, width, height (*lwh*).
Formula for volume: $V = lwh$
Volume is length × width × height.

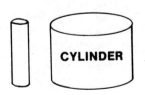

Cylinder
Dimensions: any dimension from which the area of the base may be found; and height.
Formula for volume: $V = Bh$
Volume is area of the base (*B*) × height.

The formula $V = Bh$ *may be used for any of the solids illustrated.*

Example A

Find the volume of an aquarium that is 2 ft. long, 12 in. wide, and 18 in. high.

STEPS

One: The shape is rectangular; $V = lwh$

Two: Dimensions: (2 by 1 by 1.5) ft.

Three: Volume is (2 × 1 × 1.5) cu. ft., or 3 cu. ft.

Example B

Find the volume of a cylindrical tank that has a diameter of 8 ft. and a length of 22 ft.

STEPS

One: The shape is a cylinder; $V = Bh$

Two: $B = 3.14 \times 4^2$, (πr^2), $h = 22$

Three: Volume is (3.14 × 16 × 22) cu. ft., or 1105.28 cu. ft.

Example C

How many photo displays in the shape of an eight-inch cube can be packed in a carton that is 2′ by 2′ by 4′?

STEPS

One: Problem involves cubes in a rectangular box.

   Volume of box ÷ volume of cube. $\dfrac{lwh}{s^3}$

Two: Change feet to inches or inches to feet.

   Dimensions of carton: (24 by 24 by 48) in.

Three: $\dfrac{\text{Volume of carton:} \overset{3}{\cancel{24}} \times \overset{3}{\cancel{24}} \times \overset{6}{\cancel{48}}}{\text{Volume of 1 cube:} \cancel{8} \times \cancel{8} \times \cancel{8}} = 54$

Answer: 54 photo displays can be packed in the carton.

Note: You could, if you wished, find each volume and then divide.

   $\dfrac{\text{Volume of carton:}}{\text{Volume of cube:}} = \dfrac{27\,648}{512} = 54$, but the solution shown is easier.

EXERCISE M – 9

1. Find the volume of each of the following cubes.

   (a) $s = 7$ in.   (b) $s = 6.2$ cm

2. Find the length of one side of a cube if

   (a) the volume is 1 cu. in.   (b) the volume is 27 cm^3

   (c) the area of one face is 64 sq. in.

3. Three cylindrical tanks are each 10 ft. high. Find their volumes if

   (a) the first has a diameter of 7 ft.

   (b) the area of the base of the second is 43.25 sq. ft.

   (c) the third is as wide as it is high.

4. Two tanks have the same capacity. One is rectangular in shape and the other is a cylinder. The area of the base in each case is 742 sq. ft. Which tank is taller?

5. Using the formula $V = lwh$, find the volumes of the rectangular solids illustrated below.

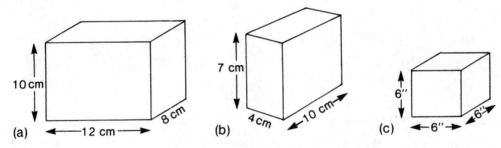

(a) ◄—— 12 cm ——► 8 cm   10 cm

(b) 7 cm   4 cm   10 cm

(c) 6″   ◄—6″—►   6″

6. Find the height of each three-dimensional figure illustrated.

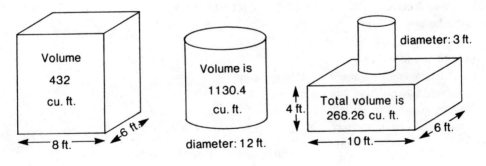

Volume 432 cu. ft.   ◄— 8 ft. —►   6 ft.

Volume is 1130.4 cu. ft.   diameter: 12 ft.

diameter: 3 ft.   4 ft.   Total volume is 268.26 cu. ft.   ◄—10 ft.—►   6 ft.

7. Using the formula $V = Bh$, and its variations $B = \dfrac{V}{h}$ and $h = \dfrac{V}{B}$, complete the following chart.

	Solid	Area of base	Height	Formula	Volume
(a)	Cylinder	325 cm²		$h = \dfrac{V}{B}$	2600 cm³
(b)	Rectangular solid		15 ft.		600 cu. ft.
(c)	Cylinder		11.2 m		1289.12 m³
(d)	Cube	64 sq. ft.			

ANSWERS FOR EXERCISE M - 9

1. (a) 343 cu. in.   (b) 238.328 cm³
2. (a) 1 in.   (b) 3 cm   (c) 8 in.
3. (a) 384.65 cu. ft.   (b) 432.5 cu. ft.   (c) 785 cu. ft.
4. same height
5. (a) 960 cm³   (b) 280 cm³   (c) 216 cu. in.
6. (a) 9 ft.   (b) 10 ft.
   (c) Total volume − Volume of rectangular box
       = 268.26 cu. ft. − 240 cu. ft.
   Volume of cylinder is 28.26 cu. ft.
   Area of base (cylinder) = 3.14 × 1.5 × 1.5, or 7.065 cu. ft.

   Height $= \dfrac{V}{B} = \dfrac{28.26}{7.065} = 4$. Total height is 4 ft. (box) + 4 ft. (cylinder).

   Total height is 8 ft.

7. (a) Height = 8 cm   (b) Formula: $B = \dfrac{V}{h} = \dfrac{600}{15}$ Area of base = 40 sq. ft.

   (c) Formula: $B = \dfrac{V}{h}$ $B = \dfrac{1289.12}{11.2} = 115.1$ Area of base is 115.1 m²

   (d) Since area of base in a cube is s^2, s is $\sqrt{64}$ or 8; height is 8 ft.
       Volume is 8^3, or 512 cu. ft.   Formula: $V = s^3$

ITEM M – 10 CONVERSION OF UNITS:
SQUARE AND CUBIC MEASURE

In converting units of square and cubic measure, be sure to use the correct conversion factor.

For example,

1 ft. = 12 in. **but** 1 sq. ft. = 144 sq. in. **and** 1 cu. ft. = 1728 cu. in.

1 yd. = 3 ft. **but** 1 sq. yd. = 9 sq. ft. **and** 1 cu. yd. = 27 cu. ft.

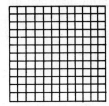

ONE SQUARE FOOT
12 in. by 12 in.
144 square inches

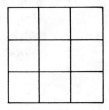

ONE SQUARE YARD
3 ft. by 3 ft.
9 square feet

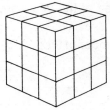

ONE CUBIC YARD
3 ft. by 3 ft. by 3 ft.
27 cubic feet

In working with area and volume, conversion of units may occur within the question or at the end of it.

Example A

Find the cost of a carpet for a floor that is 10 ft. by 18 ft. if the cost is $13.00 a square yard.	
Solution 1	*Solution 2*
Change the dimensions to yards before you find the area.	Find the area in square feet and then convert to square yards.
Dimensions in feet: 10 by 18	Dimensions in feet: 10 by 18
Dimensions in yards: $\frac{10}{3}$ by $\frac{\overset{2}{\cancel{18}}}{\cancel{3}}$	Area in square feet: 180
	Area in square yards: 180 ÷ 9 = 20
Area in square yards: 20	Cost = 20 × $13.00 = $260.00
Cost: 20 × $13.00 or $260.00	

Example B

How many tiles, each 9 inches square, would be needed to cover a floor that is 10 ft. by 18 ft.?

Solution in several steps	*Solution in one step*
Area of floor in square feet: 180	$$\frac{10 \times \overset{2}{\cancel{18}} \times \overset{16}{\cancel{144}}}{\cancel{9} \times \cancel{9}} = 320$$
Area of tile in square inches: 81	
Area of tile in square feet: $\dfrac{81}{144}$	In this solution you put down all the information in a single expression, then use cancellation to find the answer.
Number of tiles: $180 \div \dfrac{81}{144}$	
$\qquad = 320$	

EXERCISE M – 10

1. Decide whether to multiply or divide, and give the conversion factor.
   (a) To change square yards to square feet, _____ by _____.
   (b) To change square feet to square inches, _____ by _____.
   (c) To change cubic feet to cubic yards, _____ by _____.
   (d) To change cubic inches to cubic feet, _____ by _____.
   (e) To change square yards to square inches, _____ by _____.

2. Change the following as indicated.
   (a) $3\frac{1}{2}$ sq. ft. to square inches
   (b) 1.75 sq. yd. to square inches
   (c) 0.2 sq. yd. to square feet
   (d) 16 cu. ft. to cubic inches
   (e) 112.5 sq. ft. to square yards
   (f) 81 cu. ft. to cubic yards

3. Find each area in the unit given.
   (a) A rectangle 9 ft. by 1.5 yd. (Answer in square yards).
   (b) A circle with diameter 18 in. (Answer in square feet).
   (c) A triangle with base 11 in. and height 2 ft. (Answer in square inches).
   (d) A 10 inch square (Answer in square feet).

4. Find each volume in the required unit.

   (a) A 15 inch cube (Answer in cubic feet).

   (b) A rectangular solid 6 ft. by 10 ft. by 2 yd. (Answer in cubic yards).

   (c) A cylinder with base area 120 sq. ft. and height 3.5 yd. (in cu. ft.)

5. How many rectangular containers can be filled from a tank of liquid with a capacity of 12.5 cu. yd. if the dimensions of the containers are 3 ft., 5 ft., and 24 in.?

ANSWERS FOR EXERCISE M – 10

1. (a) Multiply by 9.   (b) Multiply by 144.   (c) Divide by 27.

   (d) Divide by 1728.   (e) Multiply by 1296.

2. (a) 504 sq. in.   (b) 2268 sq. in.   (c) 1.8 sq. ft.

   (d) 27 648 cu. in.   (e) 12.5 sq. yd.   (f) 3 cu. yd.

3. (a) $(9 \div 3) \times 1.5$ Answer: 4.5 sq. yd.

   (b) The radius in feet is $\frac{9}{12}$ or 0.75; $3.14 \times 0.75^2 = 1.76625$

   Area is 1.766 sq. ft.

   (c) Area is $\dfrac{11 \times 24}{2} = 132$ sq. in.   (d) Area in sq. ft. $= \dfrac{10 \times 10}{144}$, or $\dfrac{25}{36}$

4. (a) Volume in cu. ft. $= \dfrac{15 \times 15 \times 15}{12 \times 12 \times 12} = 1\frac{61}{64}$ cu. ft.

   (b) $\dfrac{6}{3} \times \dfrac{10}{3} \times 2$, or $\dfrac{6 \times 10 \times 6}{27} = 13\frac{1}{3}$; Volume $= 13\frac{1}{3}$ cu. yd.

   (c) $120 \times 10.5 = 1260$; Volume is 1260 cu. ft.

5. Use cancellation. $\dfrac{\text{Vol. of tank in cu. ft.}}{\text{Vol. of container in cu. ft.}} = \dfrac{12.5 \times 27}{3 \times 5 \times 2} = 11.25$

   11 containers can be filled from this tank.

ITEM M – 11   THE METRIC SYSTEM

Most countries of the world now use the metric system of measurement. Canada and the United States are in the process of converting from the British (or Imperial) system to the metric system. At the present time, both systems are in use and will continue to be until gradually the metric system will supplant the other. Elementary and secondary schools in Canada and the United States now teach the metric system either with or without reference to the British system.

HISTORY

The metric system was adopted for use by France in the late 1700s, but it was 1840 before it became the only and official system in that country. Ten years later, four other European countries had followed suit.

At the beginning of the century, thirty-five nations were using the system, and by 1970 almost every country in the world had converted or were in the process of conversion.

THE PRESENT SYSTEM

In 1875, an organization was formed called The International Bureau of Weights and Measures. Its function was to make changes in the metric system as needed. From time to time the system has been revised, leading to the present version. It is called *le Systeme Internationale d'unites* (The International System of Units) or SI for short.

SOME ADVANTAGES OF SI

1. It has been adopted world wide (almost), making it easier for countries to trade, to share technological happenings, and to communicate in a rapidly shrinking world.

2. Since it is a decimal system, conversion from one unit to another is accomplished by multiplying or dividing by a power of 10, a calculation which often requires nothing more than moving a decimal point.

3. The same prefixes are applied to different units of measurement — whether capacity, length, mass, or others. For example, a centimetre is one one-hundredth of a metre; a centigram is one one-hundredth of a gram; a centilitre is one one-hundredth of a litre. This makes the system much easier to learn than the Imperial system with its scores or unrelated terms.

4. The units of measure are interrelated. Cubic metres are related to kilograms, related to litres, etc. This provides a cohesive factor which is missing in other systems.

COMMON UNITS OF SI MEASUREMENTS

There are seven basic units in SI. These measure length, mass, time, electric current, temperature, luminous intensity, and amount of substance. There are also supplementary units and numerous derived units. Many of these are used for advanced technological or scientific calculation and have little relevance for the average person. The units that are examined in the following section are the more common ones likely to be encountered by all of us in the course of daily living.

1. **Metre**   The metre is used to measure length or distance. Dress materials, formerly known as yard goods, are now sold by the metre. A sheet of paper is measured in centimetres rather than in inches. Highway distances are posted in kilometres.

2. **Gram**   The gram is a measure of weight or mass. (SI prefers the use of the word "mass"; in fact, "weight" is not mentioned in SI.) Your doctor records your mass (weight) in kilograms. Chocolate bars are weighed in grams. Butter and cheese will be (or are now) sold in 250 g or 500 g packages.

3. **Litre**   The litre measures liquid capacity or volume. Milk and gasoline are sold by the litre, which is a little less than a quart. A litre is equal to a cubic decimetre. A millilitre has the same volume as a cubic centimetre which, in turn, is roughly the size of a sugar cube.

4. **Degree Celsius**   The degree Celsius measures temperature. It used to be called the degree centigrade. Centigrade means "divided into a hundred parts." The Celsius scale is divided into 100 degrees between the freezing point and boiling point of water. (0°C is the freezing point and 100°C is the boiling point.) Normal body temperature is 37°C. Recommended room temperature is 20°C − 21°C. A summer day that is over 30°C in Canada is considered hot.

5. **Second**   The second is the basic unit of time. Units of time do not conform to the decimal system but are nevertheless a part of the metric system. (Conversion factors in time include 60, 24, 7, 30 etc.) The method of measuring time is basically the same as the British system. Symbols are different however, as can be seen from the following examples.

$$60 \text{ s} = 1 \text{ min} \qquad 60 \text{ min} = 1 \text{ h} \qquad 24 \text{ h} = 1 \text{ d}$$

Numeric dating and the 24-hour clock are recommended but are not obligatory under SI. Numeric dating is year followed by month followed by day (1981 09 01 is September 1, 1981, as is 1981-09-01). The 24-hour clock numbers the hours consecutively from midnight rather than starting over again after 12 noon each day. Instead of 4:15 p.m., the 24-hour clock would read 16:15.

ITEM M – 12 PREFIXES AND MEANINGS

There are six prefixes to memorize. They apply not only to linear measure, but to mass and capacity as well.

The prefixes, their meanings and uses are given below.

Prefix	Meaning	Measure of length	Measure of mass	Measure of capacity
milli	1 one-thousandth or 0.001	millimetre mm 0.001 metre	milligram mg 0.001 gram	millitre mL 0.001 litre
centi	1 one-hundredth or 0.01	centimetre cm 0.01 metre	centigram cg 0.01 gram	centilitre cL 0.01 litre
deci	one tenth 0.1	decimetre dm 0.1 metre	decigram dg 0.1 gram	decilitre dL 0.1 litre

The first three prefixes indicate fractions or parts of the basic unit. The next three indicate multiples of the basic unit.

Prefix	Meaning	Measure of length	Measure of mass	Measure of capacity
deca	10	decametre dam 10 metres	decagram dag 10 grams	decalitre daL 10 litres
hecto	100	hectometre hm 100 metres	hectogram hg 100 grams	hectolitre hL 100 litres
kilo	1000	kilometre km 1000 metres	kilogram kg 1000 grams	kilolitre kL 1000 litres

TO HELP YOU REMEMBER:

milli — Think of tax rates; 1 mil $= \dfrac{1}{1000}$ of a dollar

centi — Think of a cent; 1 cent $= \dfrac{1}{100}$ of a dollar

deci — Think of a dime; 1 dime $= \dfrac{1}{10}$ of a dollar

deca — Think of a decade; 1 decade $= 10$ years

The most commonly used prefixes are milli, centi, and kilo. The others are rarely used.

There are two other prefixes that you should recognize, one large and one very small.

Mega means one million; the symbol is M. Mg = megagram, or one million grams.

Micro means one millionth; the symbol is μ. μg = microgram or 0.000 001 g.

EXERCISE M – 12

1. Give the name and the symbol of the metric unit of measurement which replaces each of the phrases below.

EXAMPLE

(i) One one-thousandth of a gram	milligram mg
(ii) Ten litres	decalitre daL

(a) One thousand metres   (b) One hundred litres

(c) One tenth of a gram   (d) One one-thousandth of a metre

(e) One million litres   (f) One millionth of a gram

(g) One hundred grams   (h) Ten litres

(i) One tenth of a litre   (j) One thousand grams

2. Write the word that matches each symbol.

(a) mm   (b) cg   (c) dL   (d) dag   (e) km   (f) mg

(g) kL   (h) μg   (i) Mg   (j) daL   (k) hm   (l) cm

3. Arrange in order from smallest to largest:

cm, km, dam, mm, hm, dm

4. Identify each of the following statements as either true or false. Correct the errors.

(a) A milligram is smaller than a centigram.

(b) It takes one hundred centimetres to make one decimetre.

(c) The symbol for kilogram is Kg.

(d) If a metre is divided into ten equal segments, each one is called a decimetre.

(e) Deci means the same as deca.

(f) A kilolitre is 1 000 000 times as large as a millilitre.

(g) All metric symbols are written with small letters.

ANSWERS FOR EXERCISE M – 12

1. (a) kilometre   km   (b) hectolitre   hL   (c) decigram   dm
   (d) millimetre   mm   (e) megalitre   ML   (f) microgram μg
   (g) hectogram   hg   (h) decalitre   daL   (i) decilitre   dL
   (j) kilogram   kg

2. (a) millimetre   (b) centigram   (c) decilitre   (d) decagram
   (e) kilometre   (f) milligram   (g) kilolitre   (h) microgram
   (i) megagram   (j) decalitre   (k) hectometre   (l) centimetre

3. mm, cm, dm, dam, hm, km

4. (a) True   (b) False; it takes ten.   (c) False; small k.
   (d) True   (e) False; deci means one tenth; deca means ten.
   (f) True   (g) False.

ITEM M – 13 CONVERSION OF METRIC UNITS

STEPS

One: *Identify the number of steps or intervals.*

Two: *Use that number as the exponent of the conversion factor.*

Three: *Multiply or divide by the conversion factor.*

Note: You *must* be familiar with the six prefixes in the correct sequence.

Example A

| milli |
| centi |
| deci |
| UNIT |
| deca |
| hecto |
| kilo |

Change 4.2 km to metres.

STEPS

One: From kilo to basic unit is 3 steps.

Two: The conversion factor is 10^3.

Three: 4.2×10^3 (move 3 places) = 4200

4.2 km = 4200 m

Example B

| milli |
| centi |
| deci |
| UNIT |
| deca |
| hecto |
| kilo |

How many milligrams are there in 0.052 kg?

STEPS

One: From kilo to milli is 6 steps.

Two: The conversion factor is 10^6.

Three: 0.052×10^6 (move 6 places) = 52 000

There are 52 000 mg in 0.052 kg.

Example C

| milli |
| centi |
| deci |
| UNIT |
| deca |
| hecto |
| kilo |

Find the number of decilitres in 5000 mL.

STEPS

One: From milli to deci is 2 steps.

Two: The conversion factor is 10^2.

Three: $5000 \div 10^2$ (move 2 places) = 50.00

There are 50 dL in 5000 mL.

EXERCISE M – 13

Complete the following table.

Change	To	Number of steps	Conversion factor	Move dec. point	Answer
0.06 dam	centimetres	3	10^3	3 right	60 cm
3000 cm	hectometres	4	10^4	4 left	0.3 hm
8.056 dag	decigrams	2	10^2		
0.014 km	metres	3			
9.75 L	millilitres				
12 000 g	hectograms				
0.875 m	millimetres				

Answers on page 160.

We will compare the British system of linear measure with the metric.

1. **The British System** The British system uses a different word for each unit of measure. From smallest to largest, the units are: the mil, the inch, the foot, the yard, the rod, and the mile.

    Using the yard as the basic unit of measurement, we will relate each of the other units to it. The chart below shows the relationships.

$\frac{1}{36000}$ yd.	$\frac{1}{36}$ yd.	$\frac{1}{3}$ yd.	$5\frac{1}{2}$ yd.	1760 yd.
1 mil	1 inch	1 foot	1 rod	1 mile

2. **The Metric System** The metric system uses the metre for the basic unit of length. It too has smaller units that are fractions of a metre, and larger units that are multiples of a metre. From smallest to largest, these units are: the millimetre, centimetre, decimetre, metre, decametre, hectometre, and kilometre.

    The chart below shows the relationships.

0.001 metre	0.01 metre	0.1 metre	10 metres	100 metres	1000 metres
1 millimetre	1 centimetre	1 decimetre	1 decametre	1 hectometre	1 kilometre

One sees the simplicity and logic of the metric system compared to the British. Note that

(a) the words are related; each contains the word "metre", modified by a meaningful prefix.

(b) the sequence is in uniform steps; each unit is ten times as large as the preceding unit.

The following example compares conversion in the British system with conversion in the metric system. The first requires tedious calculation with unwieldy numbers, while the latter requires nothing more than the moving of a decimal point.

THE BRITISH SYSTEM	ANSWER
To convert 5.2 yards to mils, multiply 5.2 by 36 000.	187 200 mils
To convert 5.2 yards to inches, multiply 5.2 by 36.	187.2 inches
To convert 5.2 yards to feet, multiply 5.2 by 3.	15.6 feet
To convert 5.2 yards to rods, divide 5.2 by $5\frac{1}{2}$.	0.945 45 rods
To convert 5.2 yards to miles, divide 5.2 by 1760.	0.002 954 5 miles

THE METRIC SYSTEM	ANSWER
To convert 5.2 metres to millimetres, multiply 5.2 by 1000.	5200 mm
To convert 5.2 metres to centimetres, multiply 5.2 by 100.	520 cm
To convert 5.2 metres to decimetres, multiply 5.2 by 10.	52 dm
To convert 5.2 metres to decametres, divide 5.2 by 10.	0.52 dam
To convert 5.2 metres to hectometres, divide 5.2 by 100.	0.052 hm
To convert 5.2 metres to kilometres, divide 5.2 by 1000.	0.0052 km

ANSWERS FOR EXERCISE M – 13

8.056 dag	decigrams	1	10^2	2 right	805.6 dg
0.014 km	metres	3	10^3	3 right	14 m
9.75 L	millilitres	3	10^3	3 right	9750 mL
12 000 g	hectograms	2	10^2	2 left	120 hg
0.875 m	millimetres	3	10^3	3 right	875 mm

ITEM M—14 STANDARDIZATION IN WRITTEN FORMAT

1. Numerals with 5 or more digits are separated into groups of three. The groups are separated by an empty space rather than a comma.
   EXAMPLES: (a) 42 614 (not 42,614) (b) 4 020 009 852 (not 4,020,009,852)

2. Pure decimal fractions (those which are not part of a mixed number) are preceded by 0 in the whole number units place.
   EXAMPLES: (a) 0.42 rather than .42 (b) 0.005 instead of .005

3. The full names of metric units of measurement may differ from one country to another, but the symbols are universal. The symbols are not abbreviations; they do not require a period at the end.
   The symbol for millimetre is mm (not mm.) for centigram cg (not cg.)

4. The symbol rather than the full name is preferred when accompanied by a numeral.
   EXAMPLES: 30 mg rather than 30 milligrams   4.5 L rather than 4.5 litres

5. In measurements denoting area and volume, the exponents 2 and 3 are used instead of the abbreviated sq. and cu. that accompany British measures.
   EXAMPLES: The area of a rectangle 5 yd. by 7 yd. is 35 sq. yd.
   BUT   The area of a rectangle 5 m by 7 m is 35 m^2

   The volume of a cube with side 7 in. is 343 cu. in.
   BUT   The volume of a cube with side 7 cm is 343 cm^3

6. Speeds such as metres per second use the word, per, if written in full, and the solidus (/) when symbols are used.
   m/s represents metres per second. km/h represents kilometres per hour. $4/kg means $4.00 per kilogram.

7. Symbols for units of time are different although the words are the same.
   EXAMPLES: 60 s = 1 min   60 min = 1 h   24 h = 1 d

EXERCISE M – 14

1. Correct the following to conform with standard SI format.
    (a) It was a distance of 325,464,227 km.
    (b) The box was .7 m in length.
    (c) He drove at a speed of 120 km per hour.
    (d) I bought 14 Kg of potatoes.
    (e) 1 mg = 1 000 000 g.
    (f) The bottle held 5000 cLs of liquid.
    (g) The area of the wall is 35 sq. m.
    (h) The capacity of the carton was 1000 cu. cm.
    (i) The medicine cost $50.00 per gram.

2. Use numeric dating to indicate the following dates.
    (a) Dec. 25, 1980   (b) Jan. 1, 1984
    (c) July 1, 1867   (d) March 21, 2000

ANSWERS FOR EXERCISE M - 14

1. (a) A space should be used instead of a comma to separate the digits. 325 464 227 km
    (b) 0.7 instead of .7
    (c) Should be distance followed by solidus followed by time. 120 km/h
    (d) Should be small k; 14 kg
    (e) Capital M on megagram; 1 Mg
    (f) Symbols for singular and plural are the same; should be cL
    (g) Should be 35 m^2
    (h) Should be 1000 cm^3
    (i) See (c); should be $50/g

2. (a) 1980 12 25   (b) 1984 01 01   (c) 1867 07 01   (d) 2000 03 21

CHAPTER SIX : **ALGEBRA**

Algebra is the branch of mathematics which investigates in a general way the relations and properties of numbers. It is sometimes called the generalization or extension of arithmetic in which symbols, usually letters of the alphabet, represent numbers.

Example: Arithmetic deals with specific numbers using symbols that are familiar such as 20, 3000, or 6.25.

Algebra deals with general numbers using symbols that are less familiar such as x and y and mr^2.

Example: Arithmetic says that the area of a *specific* rectangle with dimensions of 6 ft. and 3 ft. is 18 sq. ft.

Algebra says that the area of *any* rectangle with dimensions l (for length) and w (for width) is lw, and expresses this relationship in the formula: $A = lw$.

ITEMS IN CHAPTER

ITEM A − 1   ARITHMETIC AND ALGEBRA: SOME SIMILARITIES

	Arithmetic	Algebra
1. *Only like numbers may be added or subtracted.*	3 cats $+ 5$ cats $= 8$ cats 2 dogs $+ 4$ dogs $= 6$ dogs But 3 cats and 2 dogs cannot be stated as a total number of cats or dogs. $3(10) + 6(10) = 9(10)$ $3(10) + 6(12)$ cannot be added in this form.	$3c + 5c = 8c$ $2d + 4d = 6d$ But $3c$ and $2d$ cannot be combined.
2. *The order of addends does not affect the sum.*	$3 + 2 + 4 = 9$ $3 + 4 + 2 = 9$ $4 + 3 + 2 = 9$ etc.	$a + b + c$ is equal to $a + c + b$ and equal to $c + a + b$ etc.
3. *The order of factors does not affect the product.*	$2 \times 5 \times 3 = 30$ $5 \times 3 \times 2 = 30$ $3 \times 2 \times 5 = 30$ etc.	$a \times b \times c$ (or abc) $=$ $b \times c \times a$ (or bca) $=$ $c \times a \times b$ (or cab) etc.
4. *The use of exponents is the same.*	$10^4 = 10 \times 10 \times 10 \times 10$ $(6 - 2)^2 = (6 - 2)(6 - 2)$	$a^4 = a \times a \times a \times a$ $(a - b)^2 = (a - b)(a - b)$
5. *Numbers may be factored.*	$24 = 2 \times 2 \times 2 \times 3$	$6a^2b = 2 \times 3 \times a \times a \times b$
6. *Fractions may be reduced to lowest terms.*	$\dfrac{15}{20} = \dfrac{15 \div 5}{20 \div 5} = \dfrac{3}{4}$	$\dfrac{4a^2}{4ab} = \dfrac{4a^2 \div 4a}{4ab \div 4a} = \dfrac{a}{b}$
7. *Cancellation is used in both.*	$\dfrac{\overset{1}{\cancel{5}}}{\underset{4}{\cancel{12}}} \times \dfrac{\overset{\overset{1}{\cancel{3}}}{\cancel{9}}}{\underset{\underset{1}{2}}{\cancel{10}}} \times \dfrac{\overset{1}{\cancel{2}}}{\underset{1}{\cancel{3}}} = \dfrac{1}{4}$	$\dfrac{\overset{1a}{\cancel{2a}}}{\underset{2}{\cancel{4b}}} \times \dfrac{\overset{1a}{\cancel{2ab}}}{\underset{3}{\cancel{6}}} = \dfrac{a^2}{6}$

ITEM A – 2 VOCABULARY OF ALGEBRA

Variable   The letter used to represent a number in algebra

Term   A number or letter, either standing alone or joined by multiplication or division with other numbers or letters, to form one unit of an expression.

Examples: a $2xy$ ptr $\dfrac{4k}{3h}$

Like Terms   Terms that have the same literal coefficient.

Examples: $4ab$ and $6ab$ $3t^2$ and $25t^2$ $9\dfrac{m}{y}$ and $6\dfrac{m}{y}$

Coefficient   May be literal or numerical. In the term $5ab$, the literal coefficient is ab, and the numerical coefficient is 5.

Expression   May consist of one term (monomial) or more than one term joined by addition or subtraction.

Monomial   An expression consisting of one term.

Examples: $3k$ 29 $150x^2y^3$

Polynomial   An expression consisting of two or more terms.

    (a) Binomial: a two-term expression.

        Examples: $2r + 3s$ $25a^2 - 36b$ $z - y$

    (b) Trinomial: a three-term expression.

        Examples: $x^2 - 2xy + y^2$ $3ab + 2bd - 4cd$

Signs of comparison   Used in arithmetic as well as algebra; they include the following:

    (a) = "is equal to" $3 \times 5 = 15$

    (b) > "is greater than" $3 \times 5 > 12$

    (c) < "is less than" $3 \times 5 < 18$

Mathematical sentence   A statement about the relative values of two quantities (using signs of comparison).

Examples: $20 > 10$ $16 \div 4 = 4$ 20 oz. < 2 lb.

Equation   A mathematical sentence which states that two quantities are equal.

Examples: $5^2 = 25$ $1 \text{ cm} = 10 \text{ mm}$ $3b + 7 = 43$

Formula   An abbreviated method of stating a rule that has to do with the relationship between two quantities

Examples: $A = lw$ (Area = length × width)

              $i = prt$ (interest = principal × rate × time)

ITEM A – 3 SIGNED NUMBERS: MEANING AND TERMINOLOGY

SOME DEFINITIONS OF NUMBERS

Natural Numbers are counting numbers.

Examples: 1, 2, 3, etc.

Whole Numbers include zero as well as the natural numbers.

Examples: 0, 1, 2, 3, etc.

Integers are positive and negative whole numbers and zero.

Examples: $^-456$, $^+27$, $^-3000$. etc.

Rational Numbers are those that can be expressed as the ratio of two integers. They may be positive or negative. They include fractions and mixed numbers.

Examples: 48, $^-65.3$, $^+1\frac{1}{2}$, $^-0.875$, $-\frac{1}{4}$, $^+10\frac{1}{2}$, etc.

Literal Numbers are used in algebra. They are letters of the alphabet which represent a number. The formula for area uses 3 literal numbers: $A = lw$.

Other examples: $i = prt$, $V = Bh$.

Signed Numbers are positive and negative numbers. They may include any of the numbers mentioned above (except 0 which doesn't have a sign) and they may include literal as well as numerical components.

Examples: $^-4$, ^+a, ^-2ab, $^+2.5$, $^-s^3$, $\dfrac{^+55k}{^-60h}$, etc.

A Signed Number is composed of two parts: a sign of quality and an absolute value.

The Sign of Quality tells whether the number is positive or negative. It is a plus or minus sign written to the left of the number. Often it is a small sign written in the upper left; or it may be the usual sign that is used to indicate addition or subtraction. If there is no sign, the number is understood to be positive.

Positive 4 may be written: $+4$ or $^+4$ or 4.

Negative 4 may be written: -4 or $^-4$.

The Absolute Value of a signed number is its numerical value. The absolute value is not affected by the sign of quality.

The absolute value of $^+3$ is 3.

The absolute value of $^-3$ is also 3.

Positive Numbers are greater than zero; they appear to the right of zero on the horizontal number line.

Negative Numbers are less than zero; they appear to the left of zero on the horizontal number line.

A Number Line is a straight vertical or horizontal line marked off in equal segments, and showing at some point the origin (zero). A number line is endless in either direction. The written representation of a number line is selected according to the data that is to be presented.

A number line

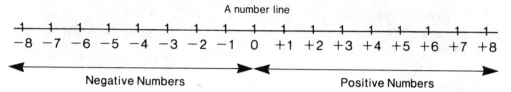

$$-8 \quad -7 \quad -6 \quad -5 \quad -4 \quad -3 \quad -2 \quad -1 \quad 0 \quad +1 \quad +2 \quad +3 \quad +4 \quad +5 \quad +6 \quad +7 \quad +8$$

Negative Numbers   Positive Numbers

A number to the right of another on the number line is considered greater in value.

$^-7$ is greater than $^-10$	$^-7 > ^-10$
$^+2$ is greater than $^-2$	$^+2 > ^-2$
0 is greater than $^-5$	$0 > ^-5$

EXERCISE A – 3

1. Write the integers only from this set:

   $12, ^+1.5, ^-600, ^-1\frac{1}{2}, ^+36b$.

2. Write the literal coefficient of each of the following:

   (a) $3a$ (b) $16b$ (c) $2m^2n^2$ (d) $\dfrac{4r}{5s}$

3. Name the numerical coefficient.

   (a) $4a^2$ (b) kb (c) $0.8\,d$

4. Use signs of comparison to make true statements of the following:

   (a) $^+9$ $^-9$ (b) $^-2$ 0 (c) $^-8.5$ $^-8$

   Note: signs of comparison include:

$=$ "is equal to"	$(3 \times 4 = 12)$
$>$ "is greater than"	$(10 > 9)$
$<$ "is less than"	$(5 < 6)$

ANSWERS FOR EXERCISE A - 3

1. $12, ^-600$

2. (a) a (b) b (c) m^2n^2 (d) $\dfrac{r}{s}$

3. (a) 4   (b) 1 (the 1 is understood)   (c) 0.8

4. (a) $^+9 > ^-9$ (b) $^-2 < 0$ (c) $^-8.5 < ^-8$

ITEM A – 4 ADDITION OF INTEGERS

There are three possibilities:	Vocabulary reminder		
1. The addends are all positive.	Addend	1 6 3	
2. The addends are all negative.	Addend	2 4 8	
3. Some addends are positive; some negative.	Sum	4 1 1	

Example A

1. The addends are all positive.	$^+24$
In this case you simply add as you would whole numbers and the result is positive.	$^+61$ $^+56$ $^+141$

Example B

2. The addends are all negative.	$^-14$
Again, you add as you would whole numbers and the answer is negative.	$^- 7$ $^-26$ $^-47$

Example C

3. Some addends are positive; some negative.

Add:

STEPS	$^-4 + {}^+12 + {}^+7 + {}^-15 + {}^+9 + {}^-3$
First: Add the positive numbers.	$^+12 + {}^+7 + {}^+9 = {}^+28$
Second: Add the negative numbers.	$^-4 + {}^-3 + {}^-15 = {}^-22$
Third: Find the difference between the sums.	The difference is 6.
Fourth: Retain the sign of the larger sum in the answer.	Since the larger sum (28) is positive, the answer will also be positive.
	Answer is $^+6$.

Example D

Find the sum of $^+12$, $^-15$, $^-40$, $^+7$, $^-52$, $^-21$, and $^+8$.

First	Second	Third	Fourth
	$^-15$		Since the larger sum is
$^+12$	$^-40$		negative, the answer is
$^+\ 7$	$^-52$	$^-128$	also negative.
$^+\ 8$	$^-21$	$^+\ 27$	
$^+27$	$^-128$	$^-101$	

EXERCISE A – 4

Find each sum. (a) $^+64 + {}^-25 + {}^-107 + {}^-36 + {}^+10 + {}^+42$

(b) $^-156 + {}^+254$

(c) $^-25 + {}^-36 + {}^-40$

(d) $24 + 35 + {}^-62 + {}^+48 + {}^-13$

(e) $^-4200 + {}^+4100$

(f) $^-1.5 + {}^-7 + {}^-0.25 + {}^+14 + {}^+0.875$

ANSWERS (a) $^-52$ (b) $^+98$ (c) $^-101$ (d) $^+32$ (e) $^-100$ (f) $^+6.125$

ITEM A – 5 SUBTRACTION OF INTEGERS
USING THE NUMBER LINE

Locate the subtrahend on the number line. This can be visualized; it doesn't require actual use of the number line. Then determine how many places you must move and in which direction in order to reach the minuend. If you use this method, you must be careful always to move from the subtrahend to the minuend, never the opposite. The following examples show how this works.

Question	Representation on a number line	Movement	Answer
$^+10 - {}^+4$	0 +4 +10	6 places right	$^+6$
$^+10 - {}^-4$	−4 0 +10	14 places right	$^+14$
$^-10 - {}^-4$	−10 −4 0	6 places left	$^-6$
$^-10 - {}^+4$	−10 0 +4	14 places left	$^-14$

$$-50 \quad -40 \quad -30 \quad -20 \quad -10 \quad 0 \quad +10 \quad +20 \quad +30 \quad +40 \quad +50$$

Use the number line above to visualize the extent and direction of movement in each of the following:

(a) Subtract $^-42$ from $^+65$; you move from $^-42$ to 0 and continue to $^+65$ (Ans. $^+107$).

(b) $^+73 - {}^-25$; move from $^-25$ to $^+73$. The result is 98 right or $^+98$.

(c) $^-48 - {}^-22$; move from $^-22$ to $^-48$. The result is _____ .

(d) From $^-60$ take $^+45$; begin at $^+45$. The result is _____ .

(e) Subtract $^-36$ from $^-28$; the subtrahend is $^-36$. The result is _____ .

(f)	(g)	(h)	(i)	(j)
$^-620$	$^+405$	$^-92$	$^-200$	$^+465$
$- {}^+250$	$- {}^+106$	$- {}^+16$	$- {}^-55$	$- {}^-100$

ANSWERS

(c) $^-26$ (d) $^-105$ (e) $^+8$ (f) $^-870$

(g) $^+299$ (h) $^-108$ (i) $^-145$ (j) $^+565$

The next section shows the method that is generally used for subtraction of signed numbers.

ITEM A – 5 SUBTRACTION OF INTEGERS

Rule to remember:

To subtract signed numbers change the sign of the subtrahend and add the resulting signed numbers.

Vocabulary Reminder

Minuend	5 6 4
Subtrahend	2 4 1
Difference	3 2 3

EXAMPLES

Subtract	Add
26	+26
14	−14
	+12

Subtract	Add
−26	−26
−14	+14
	−12

Subtract	Add
+26	+26
−14	+14
	+40

Subtract	Add
−26	−26
+14	−14
	−40

Original form	Changed form	Answer
Subtract 2 from 6	Add −2 and 6	+4
Subtract −7 from −10	Add +7 and −10	−3
Subtract −4 from +8	Add +4 and +8	+12
23 − +17	23 + −17	+6
+23 − −17	23 + +17	+40

Note that the sign of the minuend never changes.

EXERCISE A – 5

1. Subtract in each of the following:

   (a) −326 (b) +214 (c) 175 (d) 580 (e) −24 (f) +1000
       −134   +643   −66   520   +42   2000

2. Subtract −85 from −280   3. Subtract −15 from 0
4. From +400 subtract −300   5. Take −35 from −20
6. If the minuend is −49 and the subtrahend is +51, what is the difference?
7. Complete the chart below:

	Minuend	Subtrahend	Difference
(a)	−400	−600	
(b)	+50	−28	
(c)	−75		−50
(d)		−64	−40

Note:
Minuend − subtrahend = difference
Minuend − difference = subtrahend
Subtrahend + difference = minuend

ANSWERS FOR EXERCISE A – 5

1. (a) −192   (b) −429   (c) +241   (d) +60   (e) −66   (f) −1000
2. −195   3. +15   4. +700   5. +15   6. −100
7. (a) +200   (b) +78   (c) −25   (d) −104

ITEM A – 6 MULTIPLICATION OF INTEGERS

Rule to remember:
in multiplying signed numbers

Vocabulary reminder

If the signs of the factors are alike, the product is positive.

If the signs of the factors are different, the product is negative.

Factor	2 4 3
Factor	7
Product	1 7 0 1

EXAMPLES

$^+6 \times {}^+8 = {}^+48$	The signs are alike (both positive); product is positive.
$^-6 \times {}^-8 = {}^+48$	The signs are alike (both negative); product is positive.
$^+6 \times {}^-8 = {}^-48$	The signs are different; the result is negative.
$^-6 \times {}^+8 = {}^-48$	The signs are different; the result is negative.

If there are 3 or more factors, they must still be multiplied two at a time, and the same rule applies.

EXAMPLE Multiply $^+7 \times {}^-3 \times {}^-2 \times {}^-4$

Solution A	Solution B	Solution C
$(^+7 \times {}^-3) \times {}^-2 \times {}^-4$	$(^+7 \times {}^-3) \times (^-2 \times {}^-4)$	$^+7 \times (^-3 \times {}^-2) \times {}^-4$
$(^-21 \times {}^-2) \times {}^-4$	$^-21 \times {}^+8$	$(^+7 \times {}^+6) \times {}^-4$
$^+42 \times {}^-4$	$^-168$	$^+42 \times {}^-4$
$^-168$		$^-168$

There is a shorter method in which you ignore the signs until the final step. Then:

If the number of negative factors is even, the answer is positive.

If the number of negative factors is odd, the answer is negative.

Solution D (for the example: $^+7 \times {}^-3 \times {}^-2 \times {}^-4$)

STEPS

First: Multiply the absolute values: $7 \times 3 \times 2 \times 4 = 168$

Second: Count the number of negative factors: (there are 3) 3 is an odd number so the answer is negative.

Answer: $^-168$

This is useful when the base of a power is a negative number.

Rule: If the exponent is an even number, the answer is positive. $(-3)^4 = {}^+81$

If the exponent is an odd number, the answer is negative. $(-3)^5 = {}^-243$

ITEM A - 7 DIVISION OF INTEGERS

Rule to remember
in dividing signed numbers

Vocabulary reminder

*If the signs (of dividend and divisor)
are alike, the quotient is positive.*

$$\frac{\text{(Dividend) } 45}{\text{(Divisor) } 15} = 3 \text{ (Quotient)}$$

or

*If the signs (of dividend and divisor)
are different, the quotient is negative.*

Dividend ÷ Divisor = Quotient
$$45 \div 15 \quad = 3$$

EXAMPLES

Like Signs	Like Signs	Different Signs	Different Signs
$\frac{^+63}{^+21} = ^+3$	$\frac{^-63}{^-21} = ^+3$	$\frac{^+63}{^-21} = ^-3$	$\frac{^-63}{^+21} = ^-3$
Quotient positive	Quotient positive	Quotient negative	Quotient negative

EXERCISE A - 6 and 7

1. Find the result of each of the following:
   (a) $^-24$ is divided by $^+0.5$. The result is multiplied by $^-10^2$.
   (b) $25(^-3 + ^-4) - 2(^-6 - ^-2)$
   (c) $7 \times ^-5 + ^-6$
   (d) $\frac{^-75}{^-25} + \frac{^+75}{^+25} + \frac{^-75}{^+25} + \frac{^+75}{^-25}$

2. Divide $^-400$ by $^-20$.
3. Divide $^-20$ by $^-400$.
4. Subtract $^-1^3$ from $^-2^4$.
5. Multiply each of the following by $^-1$.

   (a) $^-456$ (b) 375 (c) $^+289$ (d) 0.056 (e) $\frac{^-9}{10}$

6. Divide each of the following by $^-0.5$.
   (a) $^-56$ (b) $^-0.5$ (c) 100 (d) $^-1.5$

ANSWERS FOR EXERCISE A – 6 and 7

1. (a) $^-4800$ (b) $^-167$ (c) $^-41$ (d) 0
2. $^+20$ 3. $^+0.05$ 4. $^+17$
5. (a) $^+456$ (b) $^-375$ (c) $^-289$ (d) $^-0.056$ (e) $\frac{^+9}{10}$ or $^+\frac{9}{10}$
6. (a) $^+112$ (b) 1 (c) $^-200$ (d) $^+3$

ITEM A – 8 CHANGING WORD PHRASES TO ALGEBRAIC EXPRESSIONS

Cover the answers at the right as you write an expression for each of the following.

1. Four times a number n. $\qquad$ $4n$

2. Half of a number n decreased by 7. $\qquad$ $\dfrac{n}{2} - 7$ or $\dfrac{1}{2}n - 7$

3. The value in cents of x dollars and x dimes. $\qquad$ $110x$ cents

4. The number of cents in y quarters and half as many dimes. $\qquad$ $30y$ cents

5. If you are r years old today,
   (a) your age ten years from now $\qquad$ $r + 10$ years
   (b) your age three years ago $\qquad$ $r - 3$ years
   (c) the age of a person who is two years less than twice your age. $\qquad$ $2r - 2$ years

6. The distance a car travels in y hours at a speed of 86 km/h. $\qquad$ $86y$ km

7. (a) The number of girls in a class composed of k boys and twice as many girls. $\qquad$ $2k$
   (b) The total number of students in the class. $\qquad$ $3k$

8. (a) One-fourth the sum of p and q and r. $\qquad$ $\dfrac{p + q + r}{4}$
   (b) Twice the product of p and q and r. $\qquad$ $2pqr$
   (c) The average of p and q and r. $\qquad$ $\dfrac{p + q + r}{3}$
   (d) The difference when p and q are subtracted from r. $\qquad$ $r - (p + q)$
   $r - p - q$
   (e) The quotient when p is divided by q times r. $\qquad$ $p \div qr$ or $\dfrac{p}{qr}$

9. You have k dollars in the bank. You deposit h dollars and later withdraw s dollars. What is your balance? $\qquad$ $k + h - s$ dollars

10. A rectangle is a yards in length and b yards in width. Find (a) its area $\qquad$ ab sq. yd.
    (b) its perimeter $\qquad$ $2(a + b)$ yd.
    (c) the difference between its length and width. $\qquad$ $(a - b)$ yd.

Note: If you have difficulty with this type of exercise, you may find that it helps to substitute an arithmetic number in place of the variable, then to analyse how the problem is worked out, and to follow the same procedure with the algebraic problem.

For example, you might substitute the number 6 for the variable y in question 4 above.

The problem is worked in both ways to show that the procedure is the same.

	Arithmetic	Algebra
The number of quarters is	6	y
The number of dimes is	$6 \div 2 = 3$	$y \div 2 = \dfrac{y}{2}$
The value of the quarters is	$6 \times 25¢ = 150¢$	$y \times 25¢ = 25y¢$
The value of the dimes is	$3 \times 10¢ = 30¢$	$\dfrac{y}{2} \times 10¢ = 5y¢$
The total value is	$180¢$	$30y¢$

EXERCISE A – 8

1. Write the number of cents in
   (a) m dollars $+ n$ dollars   (b) $2k$ dollars $+ k$ dimes
   (c) $\dfrac{p}{2}$ nickels

2. Find the speed of a car that travels
   (a) 420 km in 4.8 h   (b) 400 km in s h   (c) y km in s h
   (Note: The metric symbol for hour is h rather than hr.)

3. How do you change feet to inches? Find the number of inches in
   (a) 3.5 ft.   (b) $3.5t$ ft.   (c) t ft.   (d) $(r + 2)$ ft.

4. The next consecutive number after 24 is 25, after 106 is 107, etc. Write the next consecutive number after each of the following:
   (a) 35   (b) f (c) $2k$ (d) $2k + 1$ (e) $2k - 4$ (f) $a - z$

5. The consecutive numbers on each side of 45 are 44 and 46.
   Write the numbers on each side of the following:
   (a) 52   (b) m (c) $6y$ (d) $6y - 1$ (e) $b + h$ (f) $b - h$

6. Subtract the product of r and s from the quotient of $w \div f$.

7. If I am q years old today and you are d years old today, what was the sum of our ages 10 years ago?

ANSWERS FOR EXERCISE A – 8

1. (a) $100(m + n)$ cents   (b) $210k$ cents   (c) $2.5p$ cents or $\dfrac{5p}{2}$ cents

2. (a) 87.5 km/h   (b) $\dfrac{400}{s}$ km/h   (c) $\dfrac{y}{s}$ km/h

3. (a) 42 in.   (b) $42t$ in.   (c) $12t$ in.   (d) $(12r + 24)$ in.

4. (a) 36 (b) $f + 1$ (c) $2k + 1$ (d) $2k + 2$ (e) $2k - 3$
    (f) $a - z + 1$

5. (a) 51 and 53 (b) $m - 1$ and $m + 1$ (c) $6y - 1$ and $6y + 1$
    (d) $6y - 2$ and $6y$ (e) $b + h - 1$ and $b + h + 1$
    (f) $b - h - 1$ and $b - h + 1$

6. $\dfrac{w}{f} - rs$

7. I was $(q - 10)$; you were $(d - 10)$. The sum was $(q + d - 20)$ yr.

ITEM A – 9 OPERATIONS WITH VARIABLES: ADDITION AND SUBTRACTION

Two kinds of numbers are used in algebra; they are arithmetic numbers such as 1, 2, 3, etc. and literal numbers such as a, b, x, etc. Literal numbers are also called variables; their value varies according to a given set of circumstances.

In an algebraic formula such as $A = lw$, the value of A varies with the values of l and w.

For example, if $l = 5$ and $w = 6$, $A = 30$; but if $l = 9$ and $w = 8$, $A = 72$.

Numbers containing variables may be added, subtracted, multiplied, and divided.

ADDITION

Only like terms can be added or subtracted. Like terms have the same literal coefficient.

    $6a$ and $7a$ are like terms but $6a$ and $7b$ are not like terms.

    $6a + 7a = 13a$; $6a$ and $7b = 6a + 7b$, and cannot be combined.

    $5rs - 2rs = 3rs$; $5rs - 3sr = 2rs$. sr and rs are like terms, although such terms are usually written in alphabetical order.

If a number has no numerical coefficient, it is understood to be 1 (one). $1ab$ is usually written ab, $1rh^2$ is usually written rh^2, etc.

EXERCISE A - 9

Find the sums or differences as indicated.

1. (a) $4m + m$ (b) $k + k + k$ (c) $6d - 4c$
   (d) $2m^3 - m^3$ (e) $a^4 + a^3$ (f) $4 + b + 3b + 2$
   (g) $9abc + 2bac + 4ab$ (h) $3(m - k) + 2(m - k)$

2. Which of the following pairs are like terms?
   (a) $3d$ and $8d$ (b) k and $25k$ (c) $2y^2$ and $2y^3$
   (d) $5m$ and $5n$ (e) abc and bca (f) abc and bc
   (g) $4h^2k^3$ and $9hk^3$ (h) f and f^2

3. Add $4x + 10y + 7xy + 22x^2 + 8y + 11xy$

4. Write like terms for each of the following:

(a) $7rst$ (b) $16a^2b^4$ (c) $25(a + b + c)$ (d) mn

ANSWERS FOR EXERCISE A - 9

1. (a) $5m$ (b) $3k$ (c) $6d - 4c$ (d) m^3 (e) $a^4 + a^3$
   (f) $6 + 4b$ (g) $11abc + 4ab$ (h) $5(m - k)$
2. Like terms: $3d$ and $8d$; k and $25k$; abc and bca.
3. $4x + 18y + 18xy + 22x^2$
4. (a) Any numerical coefficient followed by rst.
   (b) Any numerical coefficient followed by a^2b^4.
   (c) Any numerical coefficient followed by $(a + b + c)$.
   (d) Any numerical coefficient followed by mn.

ITEM A – 10   OPERATIONS WITH VARIABLES: MULTIPLICATION

SOME THINGS TO REMEMBER

1. Multiplication may be indicated in several different ways:

(a) The times sign	$3 \times 4 = 12$	$a \times b = ab$
(b) Brackets	$3(4) = 12$	$a(b) = ab$
	$(3)(4) = 12$	$(a)(b) = ab$
(c) Raised dot	$3 \cdot 4 = 12$	$a \cdot b = ab$

   (d) Juxtaposition (placing of numbers side by side) can be used only with literal factors, not with numerical factors.

       3×4 cannot be written as 34, but $a \times b$ can be written as ab.

2. In multiplying numbers with both literal and numerical coefficients, the numerical product is written first, then the literal cofficient in alphabetical order.

       $7hk \times 4a \times 2bk \times 5 = 280abhk^2$ $2w \times 4mw \times 0.5w = 4mw^3$

3. Any number multiplied by 0 is equal to 0.

$$15 \times 0 = 0 \qquad 2 \times 4 \times 8 \times 10 \times 0 = 0 \qquad b \times r \times 0 \times 6 = 0$$

4. If a bracket contains more than one term and is preceded by a number, each term within the bracket must be multiplied by that number.

$$5(6 + 2) = 5 \times 6 + 5 \times 2 = 30 + 10 = 40$$
$$5(4a + 2b) = 5 \times 4a + 5 \times 2b = 20a + 10b$$
$$2(2x - 6y - 3z) = 4x - 12y - 6z$$
$$3m(2 + 5m - 7n) = 6m + 15m^2 - 21mn$$

EXERCISE A – 10

1. Show 3 ways to indicate that $4m$ is to be multiplied by $5n$.
2. Write the products for the following:

   (a) $f \times k$ (b) $5m(3n)$ (c) $2a(3b + 4c - 6d)$

   (d) $0.5kp \times 0.2pk$ (e) $3x^2y \times 4xy^2$ (f) $7 \times f \times 3f \times 9a$

3. For each pair of numbers, write (i) the sum, and (ii) the product.

   (a) a and b (b) k and k (c) $3b$ and $4c$ (d) m^2 and n^2

4. Find the products.

   (a) $6s \times 2r \times 0 \times 7sr$ (b) $6s \times 2r \times 1 \times 7sr$

ANSWERS FOR EXERCISE A – 10

1. $4m \times 5n$, $4m(5n)$, $(4m)(5n)$, $4m \cdot 5n$, $(4m)5n$
2. (a) fk (b) $15mn$ (c) $6ab + 8ac - 12ad$ (d) $0.1k^2p^2$
   (e) $12x^3y^3$ (f) $189af^2$
3. (a) sum is $a + b$; product is ab (b) sum is $2k$; product is k^2
   (c) sum is $3b + 4c$; product is $12bc$
   (d) sum is $m^2 + n^2$; product is m^2n^2
4. (a) 0   (b) $84r^2s^2$

ITEM A – 11   OPERATIONS WITH VARIABLES: DIVISION

SOME THINGS TO REMEMBER

1. Division may be expressed in different ways:

    30 divided by $6 = 30 \div 6$, or $6\overline{)30}$, or $\dfrac{30}{6}$

    $3a$ divided by $4b = 3a \div 4b$, or $\dfrac{3a}{4b}$; you would rarely see $4b\overline{)3a}$

2. When algebraic division is written in the form of a fraction, it should be reduced in the same way that other fractions are reduced.

    $$\frac{35}{40} = \frac{35 \div 5}{40 \div 5} = \frac{7}{8} \qquad \frac{35a}{40b} = \frac{35a \div 5}{40b \div 5} = \frac{7a}{8b}$$

    $$\frac{35k}{40k} = \frac{35k \div 5k}{40k \div 5k} = \frac{7}{8}$$

3. It may be necessary to factor the numerator and denominator in order to reduce to lowest terms. This makes it easier to cancel.

    $$\frac{12c^2d^3}{18dc} = \frac{2 \cdot 2 \cdot \cancel{3} \cdot c \cdot \cancel{c} \cdot \cancel{d} \cdot d \cdot d}{2 \cdot \cancel{3} \cdot 3 \cdot \cancel{d} \cdot \cancel{c}} = \frac{2cdd}{3} = \frac{2cd^2}{3}$$

    $$\frac{16r^3s^4t^2}{48r^4s^5t} = \frac{\overset{1}{\cancel{16}} \cdot \cancel{r} \cdot \cancel{r} \cdot \cancel{r} \cdot \cancel{s} \cdot \cancel{s} \cdot \cancel{s} \cdot \cancel{s} \cdot \cancel{t} \cdot t}{\underset{3}{\cancel{48}} \cdot \cancel{r} \cdot \cancel{r} \cdot \cancel{r} \cdot r \cdot \cancel{s} \cdot \cancel{s} \cdot \cancel{s} \cdot \cancel{s} \cdot s \cdot \cancel{t}} = \frac{t}{3rs}$$

    If the numerator contains more than one term, each term must be divided.

    $$\frac{4a - 10b + 6c}{2} = \frac{4a}{2} - \frac{10b}{2} + \frac{6c}{2} = 2a - 5b + 3c$$

EXERCISE A – 11

1. Write two algebraic expressions, one using the division sign and the other in the form of a fraction, to indicate

    (a) m is to be divided by n

    (b) the sum of $2r$ and $6s$ is divided by the product of 3 and k

    (c) 8 is divided by the sum of $6h$ and $4k$.

2. Reduce each of the following to lowest terms.

    (a) $\dfrac{20m}{4}$ (b) $\dfrac{36ab}{54a^2}$ (c) $\dfrac{15x^2y}{25xy^7}$ (d) $\dfrac{3(a + b)}{4a(a + b)}$

    (e) $\dfrac{0.6m^3}{0.2m}$ (f) $\dfrac{21j^2k^4l^3}{51j^5k^2l}$

3. Write the quotients in simplest form.

   (a) $42b^5 \div 6$ (b) $18d \div 16d$ (c) $16m^4n \div 4mn^2$

4. Write an expression in which the dividend is the sum of m^2 and n^2, and the divisor is the product of m and n. What is the quotient?

ANSWERS FOR EXERCISE A – 11

1. (a) $m \div n$ or $\dfrac{m}{n}$ (b) $(2r + 6s) \div (3 \times k)$ or $\dfrac{2r + 6s}{3 \times k}$

   (c) $8 \div (6h + 4k)$ or $\dfrac{8}{6h + 4k}$

2. (a) $\dfrac{20m}{4} = 5m$ (b) $\dfrac{36ab}{54a^2} = \dfrac{18 \cdot 2 \cdot a \cdot b}{18 \cdot 3 \cdot a \cdot a} = \dfrac{2b}{3a}$

   (c) $\dfrac{15x^2y}{25xy^7} = \dfrac{3 \cdot \not{5} \cdot x \cdot \not{x} \cdot \not{y}}{5 \cdot \not{5} \cdot \not{x} \cdot \not{y} \cdot y \cdot y \cdot y \cdot y \cdot y \cdot y} = \dfrac{3x}{5y^6}$

   (d) $\dfrac{3(a + b)}{4a(a + b)} = \dfrac{3}{4a}$ (e) $\dfrac{0.6m^3}{0.2m} = 3m^2$ (f) $\dfrac{\not{3} \cdot 7 \cdot \not{j} \cdot \not{j} \cdot \not{k} \cdot \not{k} \cdot k \cdot k \cdot \not{l} \cdot l \cdot l}{\not{3} \cdot 17 \cdot \not{j} \cdot \not{j} \cdot j \cdot j \cdot \not{k} \cdot \not{k} \cdot \not{l}} = \dfrac{7k^2l^2}{17j^3}$

3. (a) $42b^5 \div 6 = 7b^5$ (b) $18d \div 16d = \frac{9}{8}$ or $1\frac{1}{8}$ or 1.125 (c) $\dfrac{4m^3}{n}$

4. $(m^2 + n^2) \div mn$, or $\dfrac{m^2 + n^2}{mn}$ The quotient is $\dfrac{m^2 + n^2}{mn}$

ITEM A – 12 OPERATIONS WITH VARIABLES: MULTIPLICATION & DIVISION OF POWERS

REMINDERS

A power is composed of a *base* and an *exponent*.

The *base* is the number that is multiplied.

The *exponent* is the number of times that the base is multiplied.

The *index* is the same as the *exponent*.

Indices is the plural of *index*.

In the power 12^5, the base is 12 and the exponent or index is 5.

12^5 means that 12 is multiplied by itself 5 times.

$12^5 = 12 \times 12 \times 12 \times 12 \times 12$, or $248\,832$.

To multiply powers that have the same base, add the indices.

$$4^2 \times 4^3 = (4 \times 4) \times (4 \times 4 \times 4) \qquad c^2 \times c^3 = (c \times c) \times (c \times c \times c)$$
$$= 4 \times 4 \times 4 \times 4 \times 4 \qquad\qquad = c \times c \times c \times c \times c$$
$$= 4^5 \qquad\qquad\qquad\qquad\qquad = c^5$$
$$4^2 \times 4^3 = 4^{2+3} \text{ or } 4^5 \qquad\qquad c^2 \times c^3 = c^{2+3} \text{ or } c^5$$

EXAMPLES

$m^3 \times m^4 \times m = m^{3+4+1} = m^8$	$(a-b)^2 \times (a-b)^3 = (a-b)^5$

To divide powers that have the same base, subtract the indices.

$$6^5 \div 6^2 = \frac{\cancel{6} \cdot \cancel{6} \cdot 6 \cdot 6 \cdot 6}{\cancel{6} \cdot \cancel{6}} = 6^3 \qquad a^5 \div a^2 = \frac{\cancel{a} \cdot \cancel{a} \cdot a \cdot a \cdot a}{\cancel{a} \cdot \cancel{a}} = a^3$$
$$6^5 \div 6^2 = 6^{5-2} \qquad = 6^3 \qquad a^5 \div a^2 = a^{5-2} \qquad = a^3$$
$$\frac{b^{10}}{b^5} = \frac{\cancel{b} \cdot \cancel{b} \cdot \cancel{b} \cdot \cancel{b} \cdot \cancel{b} \cdot b \cdot b \cdot b \cdot b \cdot b}{\cancel{b} \cdot \cancel{b} \cdot \cancel{b} \cdot \cancel{b} \cdot \cancel{b}} = b^{10-5} = b^5$$

Note: Addition and subtraction of indices is possible only when the base is the same.

$7^4 \times 7^2 = 7^6$, but $7^4 \times 6^2$ does not meet the conditions.

$a^8 \div a^5 = a^3$, but $a^8 \div b^5$ cannot be simplified.

EXERCISE A -- 12

Write the appropriate products or quotients.

(a) $a^7 \times a^4$ (b) $b^2 \times b^5 \times b^3$ (c) $c^{10} \div c^8$

ANSWERS FOR EXERCISE A -- 12

(a) a^{11} (b) b^{10} (c) c^2

A-13 OPERATIONS WITH SIGNED NUMBERS

All the rules that govern operations with integers also apply in algebra.

RULE	INTEGERS	ALGEBRA
Addition with signed numbers 1. Add the positive numbers. 2. Add the negative numbers. 3. Keep the sign of the larger.	Add: $^+24, ^-50, ^-72, ^+15$ $\begin{array}{r} ^+24 \\ + ^+15 \\ \hline ^+39 \end{array} \qquad \begin{array}{r} ^-50 \\ - ^-72 \\ \hline ^-122 \end{array}$ $^+39 + ^-122 = ^-83$	Add: $2a, ^-6a, ^-15a, ^+3a$ $\begin{array}{r} ^+2a \\ + ^+3a \\ \hline ^+5a \end{array} \qquad \begin{array}{r} ^-6a \\ + ^-15a \\ \hline ^-21a \end{array}$ $^+5a + ^-21a = ^-16a$
Subtraction of signed numbers 1. Change the sign of the subtrahend. 2. Add the resulting signed numbers.	$\begin{array}{r} ^-56 \\ - ^-22 \\ \hline \end{array}$ becomes $\begin{array}{r} ^-56 \\ + ^+22 \\ \hline ^-34 \end{array}$ $\begin{array}{r} ^+48 \\ - ^+16 \\ \hline \end{array}$ becomes $\begin{array}{r} ^+48 \\ + ^-16 \\ \hline ^+32 \end{array}$	$\begin{array}{r} ^+7a \\ - ^-4a \\ \hline \end{array}$ becomes $\begin{array}{r} ^+7a \\ + ^+4a \\ \hline ^+11a \end{array}$ $\begin{array}{r} ^-a \\ - ^-6a \\ \hline \end{array}$ becomes $\begin{array}{r} ^-a \\ + ^+6a \\ \hline ^+5a \end{array}$
Multiplication of signed numbers 1. If the signs are alike, the product is positive. 2. If the signs are different, the product is negative. 3. If there is an odd number of negative factors, the product is negative.	$^-10 \times ^-6 = ^+60$ $^+4 \times ^+18 = ^+72$ $^-7 \times ^-4 \times ^+1 = ^+28$ $^-4 \times ^+16 = ^-64$ $^-2 \times ^-3 \times ^-1 \times ^+4 \times ^-5$ (4 negative factors) Product is positive: $^+120$	$^-2a \times ^-5a = ^+10a^2$ $^+3a \times ^+9b = ^+27ab$ $^-2a \times ^+5a \times ^+4b = ^-40a^2b$ $^+a \times ^-b = ^-ab$ $^-a \times ^-b \times ^+2 \times ^+3 \times ^-4$ (3 negative factors) Product is negative: ^-24ab
Division of signed numbers 1. If the signs are alike, the quotient is positive. 2. If the signs are different, the quotient is negative.	$\dfrac{^-36}{^-12} = ^+3 \qquad \dfrac{^+36}{^+12} = ^+3$ $\dfrac{^-36}{^+12} = ^-3 \qquad \dfrac{^+36}{^-12} = ^-3$	$\dfrac{^-4a^2b}{^-2b} = ^+2a^2 \qquad \dfrac{^+abc}{^+ac} = ^+b$ $\dfrac{^-6acd}{^+8ad} = -\frac{3}{4}c \qquad \dfrac{^+12a}{^-4a} = ^-3$
Use of exponents 1. If the base is negative and the exponent is an odd number, the result is negative.	$^-10^2 = ^+100$ (2 is even) $^-10^3 = ^-1000$ (3 is odd) $^-2^5 = ^-32$ (5 is odd)	$(^-4a)^2 = ^+16a^2$ $(^-4a)^3 = ^-64a^3$ $(^-2a)^5 = ^-32a^5$

The next few pages deal with each of the operations separately.

ITEM A – 14 OPERATIONS WITH SIGNED NUMBERS:
ADDITION & SUBTRACTION

ADDITION

1. $5a - b + 8c$ 2. $7x - 15$ 3. $2m + 15n$ 4. $8(a + b)$ 5. ^-4rs
    $a + 6b - 3c$ $^+4x - 22$ $^-3m + n$ $^-4(a + b)$ ^+4rs
    $\overline{7a - b + c}$ $\overline{^-10x - 18}$ $\overline{4m - 32n}$ $\overline{^-5(a + b)}$

SUBTRACTION

Remember to change the sign of the subtrahend, then add.

1. $7(k + l)$ 2. $6a + 2b$ 3. $^-31d - 45e - 24f$ 4. $2y - z$
    $\overline{^-4(k + l)}$ $\overline{3a - 9b}$ $\overline{^-30d - 50e + 24f}$ $\overline{^-6y + 4}$

REMOVAL OF BRACKETS

If a bracket is preceded by a $+$ sign or a positive number, the signs within the bracket do not change when the bracket is removed.

If a bracket is preceded by a $-$ sign or a negative number, each sign within the bracket must be changed to the opposite when the bracket is removed.

$$^+2(a + b) - 3(k - l) = {}^+2a + 2b - 3k + 3l$$
$$3(4a + 2b) - 2(7a - 2b) = 12a + 6b - 14a + 4b = {}^-2a + 10b$$

Remove brackets and collect like terms.

1. $5(2h - 3b) - 4(^-3h + 8b)$ 2. $^-7(^-3x + 2y) - 8(^-2x - 4y)$
3. $a^2 - b^2 - (5ab - 3n^2)$ 4. $(r - s) - (2r - 2s)$
5. $-(j + k - l) - 2(j - k + l)$ 6. $3b(b + 4) - 2b^2 - 4b(3b - 5)$

SUBTRACTION INVOLVING ZERO

When 0 is subtracted from any number, the result is unchanged.

$$25 - 0 = 25 \qquad ^-25 - 0 = {}^-25 \qquad (a - b) - 0 = a - b$$

When a signed number is subtracted from 0, the answer is the same number with the opposite sign.

$0 - 25 = {}^-25$ $\qquad\qquad$ $0 - a = {}^-a$
$0 - {}^-25 = {}^+25$ $\qquad\qquad$ $0 - (^-a) = {}^+a$
$0 - (a + b - c) = {}^-a - b + c$ $\qquad$ $0 - (2a - 4b + 6c) = {}^-2a + 4b - 6c$

1. Subtract $(^-2m + 3n - 12)$ from 0.   2. Subtract $(1 + 5 - 8)$ from 0.
3. $0 - (^-2v + 3w - 6)$ 4. $0 - {}^-5$

ANSWERS

Addition	Subtraction	Removal of Brackets	With Zero
1. $13a + 4b + 6c$	1. $11(k + l)$	1. $22h - 47b$	1. $2m - 3n + 12$
2. $x - 55$	2. $3a + 11b$	2. $37x + 18y$	2. $^+2$
3. $3m - 16n$	3. $^-d + 5e - 48f$	3. $a^2 - b^2 - 5ab + 3n^2$	3. $2v - 3w + 6$
4. $^-1(a + b)$	4. $8y - z - 4$	4. $^-r + s$	4. $^+5$
5. 0		5. $^-3j + k - l$	
		6. $^-11b^2 + 32b$	

ITEM A – 15 MULTIPLICATION WITH SIGNED NUMBERS

REMINDER

If the signs are alike, the answer is positive.

If the signs are different, the result is negative.

$$5 \times 6 = 30 \quad (-5)(-6) = 30 \quad ^-5(6) = ^-30 \quad 5(-6) = ^-30$$
$$a \times v = av \quad (^-a)(-v) = av \quad ^-a(v) = ^-av \quad a(-v) = ^-av$$

EXERCISE A – 15

1. $(-1)(-1)(-1)$
2. $4a(-2b)$
3. $20(-4a) \times 3$
4. $0.2y \times 0.4y^2$
5. $^-r(r^3 - r^2)$
6. $^-x(x - 1)$
7. $^-2b \times 4a^2c \times 0$
8. $^-ab \times a(-bc)$
9. $7d^2e^3f^4 \times (-1)$
10. $(-6mn)(5m^2n^3)$
11. $0.05d(0.4cd)$

REMINDER

If the number of negative factors is even, the product is positive.

If the number of negative factors is odd, the product is negative.

12. Tell whether the answers for the following will be positive or negative.

(a) $(-3)^3$ (b) $(-5)^6$ (c) $^-9(10)$ (d) $(-a)^3 \times a \times (-a)^5$

Study this example.	$a^3 + 5a^2 - 3a + 7$
	$\times\ 2a\ -\ 4$
Multiply each term by ⁻4.	$-4a^3 - 20a^2 + 12a - 28$
Multiply each term by $2a$.	$2a^4 + 10a^3 - 6a^2 + 14a$
Add the products.	$2a^4 + 6a^3 - 26a^2 + 26a - 28$

13. $4a^3 - 3a^2 - 5a + 4$
$\underline{\times\ 2a - 3}$

14. $5a^2 - 4a + 7$
$\underline{\times\ 3a^2 + a - 2}$

15. $a^3 - 3a^2b + 3ab^2 - b^3$
$\underline{\times\ a^2 - 2ab + b^2}$

Study this example	$a + 5$	$c - d$	$4b + 2k$
	$\underline{a + 3}$	$\underline{c - d}$	$\underline{3b - k}$
Multiply each term by 3.	$3a + 15$	$-cd + d^2$	$-4bk - 2k^2$
Multiply each term by a.	$\underline{a^2 + 5a}$	$\underline{c^2 - cd}$	$\underline{12b^2 + 6bk}$
Add the products.	$a^2 + 8a + 15$	$c^2 - 2cd + d^2$	$12b^2 + 2bk - 2k^2$

16. $b + 6$
$\underline{b - 3}$

17. $1 - h$
$\underline{1 - h}$

18. $3r - 4t$
$\underline{^-5r + t}$

19. $^-2m - 2n$
$\underline{^-2m - 2n}$

20. $a - b$
$\underline{b - a}$

ANSWERS FOR EXERCISE A - 15

1. $^-1$ 2. ^-8ab 3. ^-240a 4. $0.08y^3$
5. $^-r^4 + r^3$ 6. $^-x^2 + x$ 7. 0 8. $^+a^2b^2c$
9. $-7d^2e^3f^4$ 10. $-30m^3n^4$ 11. $0.02cd^2$
12. (a) negative   (b) positive   (c) negative   (d) positive
13. $8a^4 - 18a^3 - a^2 + 23a - 12$ 14. $15a^4 - 7a^3 + 7a^2 + 15a - 14$
15. $a^5 - 5a^4b + 10a^3b^2 - 10a^2b^3 + 5ab^4 - b^5$
16. $b^2 + 3b - 18$ 17. $1 - 2h + h^2$ 18. $^-15r^2 + 23rt - 4t^2$
19. $4m^2 + 8mn + 4n^2$ 20. $^-a^2 + 2ab - b^2$

ITEM A – 16 DIVISION WITH SIGNED NUMBERS

REMINDER

If the signs are alike, the quotient is positive.

If the signs are different, the quotient is negative.

EXAMPLES

$$\frac{40}{10}=4 \quad \frac{^-40}{^-10}=^+4 \quad \frac{40}{^-10}=^-4 \quad \frac{^-40}{^+10}=^-4 \quad \frac{^-k}{^-k}=^+1 \quad \frac{k}{^-k}=^-1$$

EXERCISE A – 16

Divide.

1. 0.16 by $^-0.02$ 2. $\frac{^-3}{4}$ by $\frac{4}{5}$ 3. $3mn$ by $\frac{1}{2m}$ 4. $^-4a^2b^3$ by $^-a^3b^2$

REMINDER

To divide by a fraction, invert the divisor, and multiply.

EXAMPLES

$$\frac{3}{4}\div\frac{5}{8}=\frac{3}{\cancel{4}}\times\frac{\cancel{8}^{2}}{5}=\frac{6}{5}=1\frac{1}{5} \qquad \frac{a}{b}\div\frac{c}{d}=\frac{a}{b}\times\frac{d}{c}=\frac{ad}{bc}$$

$$^-24a^2\div\frac{4}{3}=^-\cancel{24}^{6}a^2\times\frac{3}{\cancel{4}}=^-18a^2$$

Divide.

5. $\frac{^-36a}{^-48ab}$ by $\frac{6b^2}{4ab}$ 6. $\frac{3a^2b}{^-4c}$ by $\frac{^-15ab}{16c}$ 7. $\frac{5mn}{2}$ by $\frac{10m^2}{15n}$

8. $5gh$ by $\frac{1}{g}$ 9. $\frac{^-14df}{7d}$ by $\frac{^-21d^2f^2}{^-2f}$ 10. $\frac{20st}{^-5s}$ by $\frac{16t}{^-4t}$

REMINDER

When dividing a polynomial by a monomial, each term of the polynomial must be divided.

EXAMPLE

$$\frac{6a^2b + 12b^3 - 24a^2b^2}{2ab} = 3a + \frac{6b^2}{a} - 12ab$$

11. $\dfrac{12m^2n^2 - 6mn}{-6mn}$ 12. $\dfrac{9f^3 - 18f^2 + 15f}{3f}$

13. $\dfrac{2.4b - 1.6b^2 + 0.8b}{-0.2b}$ 14. $\dfrac{a + ab + a^2b^2}{a}$

15. $\dfrac{6a^3b^2c - 4ab^2c^3}{-abc}$ 16. $\dfrac{6d^2 - d}{-d}$

ANSWERS FOR EXERCISE A -- 16

1. $^-8$ 2. $-\dfrac{15}{16}$ 3. $6m^2n$ 4. $\dfrac{^-4 \cdot \cancel{a} \cdot \cancel{a} \cdot b \cdot b \cdot b}{^-\cancel{a} \cdot \cancel{a} \cdot a \cdot \cancel{b} \cdot \cancel{b}} = \dfrac{^-4b}{^-a} = \dfrac{4b}{a}$

5. $\dfrac{\overset{}{\cancel{6} \cdot \cancel{6} \cdot \cancel{a}} \times \cancel{4} \cdot a \cdot \cancel{b}}{^-\underset{2}{\cancel{12}} \cdot \cancel{4} \cdot \cancel{a} \cdot b \times \cancel{6} \cdot \cancel{b} \cdot b} = \dfrac{a}{2b^2}$ 6. $\dfrac{\cancel{3} \cdot \cancel{a} \cdot a \cdot \cancel{b} \cdot 4 \cdot \cancel{4} \cdot \cancel{c}}{^-\cancel{4} \cdot \cancel{a} \cdot ^-5 \cdot \cancel{3} \cdot \cancel{a} \cdot \cancel{b}} = \dfrac{4a}{5}$

7. $\dfrac{\cancel{5m} n \times 15n}{2 \times \underset{2}{\cancel{10}} \times \cancel{m} \times m} = \dfrac{15n^2}{4m}$ 8. $5g^2h$

9. $\dfrac{^-\cancel{7} \cdot 2 \cdot \cancel{d} \cdot \cancel{f}}{7 \cdot \cancel{d}} \times \dfrac{-2 \cdot \cancel{f}}{^-\cancel{7} \cdot 3 \cdot d \cdot d \cdot \cancel{f} \cdot \cancel{f}} = \dfrac{^-4}{21d^2}$

10. $\dfrac{\cancel{4} \cdot \cancel{5} \cdot \cancel{8} \cdot t \cdot ^-\cancel{4} \cdot \cancel{t}}{^-\cancel{5} \cdot \cancel{8} \cdot \cancel{4} \cdot \cancel{4} \cdot \cancel{t}} = t$ 11. $^-2mn + 1$

12. $3f^2 - 6f + 5$ 13. $^-12 + 8b - 4 = 8b - 16$

14. $1 + b + ab^2$ 15. $^-6a^2b + 4bc^2$ 16. $^-6d + 1$

ITEM A – 17 SUBSTITUTION IN AN ALGEBRAIC EXPRESSION

STEPS

One: Copy the expression.

Two: Re-copy, substituting the numerical values with their correct signs. **Don't omit this step!**

EXAMPLES

If $a = 3$, $b = {}^-2$, and $c = {}^-1$, find the value of each expression.

1. $a + b$	2. ab^2	3. $a - bc$	4. $\dfrac{a}{bc} \times c^3$
$= 3 + ({}^-2)$	$= 3({}^-2)^2$	$= 3 - ({}^-2)({}^-1)$	
$= 1$	$= 3(4)$	$= 3 - ({}^+2)$	$= \dfrac{3}{({}^-2)({}^-1)} \times ({}^-1)^3$
	$= 12$	$= 1$	$= \dfrac{3}{2} \times ({}^-1)$
			$= {}^-1\dfrac{1}{2}$

EXERCISE A – 17

1. If $a = {}^-2$, find the value of

    (a) $6a$ (b) $\dfrac{10a}{2a}$ (c) a^2 (d) a^3 (e) ^-a

2. If $h = 3$ and $k = {}^-3$, find the value of

    (a) $h + k$ (b) $h - k$ (c) $k - h$ (d) hk

    (e) $\dfrac{h^2 k^2}{2k}$ (f) $4(k - h) + \dfrac{h}{k}$

3. If $x = 5$ and $y = {}^-3$, find the value of

    (a) $2x^2 + 3x - y^2 + 5xy$ (b) $\dfrac{3x}{2} \times \dfrac{6y}{5}$

    (c) $3x + 5y - 2y + 4x - 7xy$

4. In the formula $A = bh$, find the value of A when

    (a) $b = 9.2$ and $h = 10$ (b) $b = 4$ and $h = 1\frac{3}{4}$

5. In the formula $A = \pi r^2$, find the value of A if $r = 10$.

6. In the formula $i = prt$, find i when

    (a) $p = 2500$, $r = 0.09$, and $t = 3$

    (b) $p = 4000$, $r = 0.1125$, and $t = 2.5$

7. In the formula $V = \pi r^2 h$, find V when $r = 20$ and $h = 15$.
8. In the formula $F = \frac{9}{5}C + 32$, find F when $C = 15$.
9. In the formula $C = \frac{5}{9}(F - 32)$, find C when $F = 77$.
10. In the formula $V = s^3$, find V when $s = 0.9$.

ANSWERS FOR EXERCISE A -- 17

1. (a) $^-12$	(b) $^+5$	(c) 4	(d) $^-8$	(e) $^+2$	
2. (a) 0	(b) 6	(c) $^-6$	(d) $^-9$	(e) $^-13.5$	(f) $^-25$
3. (a) $^-19$	(b) $^-27$	(c) 131	4. (a) 92	(b) 7	
5. 314	6. (a) 675	(b) 1125	7. 18 840		
8. 59	9. 25	10. 0.729			

ITEM A -- 18 EQUATIONS

An equation is a mathematical sentence.

Mathematical sentences make statements about the relative values of two quantities.

EXAMPLES OF MATHEMATICAL SENTENCES

(a) $6 \times 4 = 24$	(6 times 4 is equal to 24)
(b) $12 \neq 13$	(12 is not equal to 13)
(c) 4 lbs. > 50 oz.	(4 lbs. is greater than 50 oz.)
(d) $6 \times 4 + 8 < 6 \times (4 + 8)$	($6 \times 4 + 8$ is less than $6 \times (4 + 8)$)
(e) $3a = 36$	(3 times a is equal to 36)

An equation is a mathematical sentence that expresses equality. Examples (a) and (e) above are equations. Example (e) is an algebraic equation. It contains an unknown element, a variable, which is represented by the letter a.

To solve an equation is to find the value of the unknown that will make the equation a true sentence.

In the equation $3a = 36$, we find the value of a that will make this statement true. We find that the value of a is 12 because 12 is the number that fits the conditions: $3 \times 12 = 36$.

THE BASIC RATIONALE IN SOLVING AN EQUATION

To begin, the two sides are equal; this equality must be maintained throughout the solution of the equation. You are free to add, subtract, multiply, or divide, using any number(s) that you choose **provided** you perform exactly the same operation on each side of the equation.

EXAMPLE

Start with the following equation:	$3\,h\,50\,min = 2\,h\,110\,min$
Add 20 min to each side.	$3\,h\,50\,min + 20\,min = 2\,h\,110\,min + 20\,min$
The 2 sides are still equal.	$4\,h\,10\,min = 4\,h\,10\,min$

STEPS IN SOLVING AN EQUATION

One: *Copy the equation and simplify if necessary.*

Two: *Identify the unwanted terms and use whatever operation is necessary to remove them.*

Three: *Prove your answer by separately substituting the value of the unknown in each side of the equation.*

EXAMPLE IN WHICH SUBTRACTION IS USED TO UNDO ADDITION

STEPS

One: Copy the equation. $\quad 2a + 3 = a + 7$

Two: Subtract 3 from each side. $\quad 2a + 3 - 3 = a + 7 - 3$

$$2a = a + 4$$

Subtract a from each side. $\quad 2a - a = a + 4 - a$

$$a = 4$$

Three:

Left Side	Right Side
$2a + 3$	$a + 7$
$= 2(4) + 3$	$= 4 + 7$
$= 8 + 3$	$= 11$
$= 11$	

EXAMPLE IN WHICH THE INVERSE OPERATIONS OF ADDITION AND DIVISION ARE USED TO UNDO SUBTRACTION AND MULTIPLICATION

STEPS

One:	Copy the equation.	$4b - 12 = 18 - 2b$
Two:	Add 12 to each side.	$4b - 12 + 12 = 18 - 2b + 12$
		$4b = 30 - 2b$
	Add $2b$ to each side.	$4b + 2b = 30 - 2b + 2b$
		$6b = 30$
	Divide each side by 6.	$\dfrac{6b}{6} = \dfrac{30}{6}$
		$b = 5$

Three:

Left Side	Right Side
$4b - 12$	$18 - 2b$
$= 4(5) - 12$	$= 18 - 2(5)$
$= 20 - 12$	$= 18 - 10$
$= 8$	$= 8$

Note: You may have learned to solve equations using the method of transposition which is an abbreviated form of the above. The same equations are solved below using this method.

$$2a + 3 = a + 7 \qquad 4b - 12 = 18 - 2b$$
$$2a - a = 7 - 3 \qquad 4b + 2b = 18 + 12$$
$$a = 4 \qquad\qquad b = 5$$

EXAMPLE USING ADDITION, SUBTRACTION, AND MULTIPLICATION

STEPS

One:	Copy the equation.	$4c - 3(c + 8) = \frac{1}{2}c + 10$
	Remove brackets.	$4c - 3c - 24 = \frac{1}{2}c + 10$
	Collect like terms.	$c - 24 = \frac{1}{2}c + 10$
Two:	Add 24 to each side.	$c - 24 + 24 = \frac{1}{2}c + 10 + 24$
		$c = \frac{1}{2}c + 34$
	Subtract $\frac{1}{2}c$ from each side.	$c - \frac{1}{2}c = \frac{1}{2}c - \frac{1}{2}c + 34$
		$\frac{1}{2}c = 34$
	Multiply each side by 2.	$\frac{1}{2}c \times 2 = 34 \times 2$
		$c = 68$

Three:

Left Side	Right Side
$4c - 3(c + 8)$	$\frac{1}{2}c + 10$
$= 4(68) - 3(68 + 8)$	$= \frac{1}{2}(68) + 10$
$= 272 - 228$	$= 34 + 10$
$= 44$	$= 44$

SOME OTHER EXAMPLES

(a) $5r - (3 + 2r) = 7$

$5r - 3 - 2r = 7$

$3r - 3 = 7$

$3r - 3 + 3 = 7 + 3$

$3r = 10$

$\dfrac{3r}{3} = \dfrac{10}{3}$

$r = 3\tfrac{1}{3}$

(b) $\dfrac{3b + 7}{4} = 7$

$\dfrac{3b + 7}{4} \times 4 = 7 \times 4$

$3b + 7 = 28$

$3b + 7 - 7 = 28 - 7$

$3b = 21$

$\dfrac{3b}{3} = \dfrac{21}{3}$

$b = 7$

(c) $0.04k - 16 = 40$

$0.04k - 16 + 16 = 40 + 16$

$0.04k = 56$

$\dfrac{0.04k}{0.04} = \dfrac{56}{0.04}$

$k = 1400$

(c) another version

$0.04k - 16 = 40$

$4k - 1600 = 4000$

$4k - 1600 + 1600 = 4000 + 1600$

$4k = 5600$

$\dfrac{4k}{4} = \dfrac{5600}{4}$

$k = 1400$

(d) $\tfrac{1}{2}y + \tfrac{1}{3}y = 5$

$\tfrac{5}{6}y = 5$

$\tfrac{5}{6}y \times 6 = 5 \times 6$

$5y = 30$

$\dfrac{5y}{5} = \dfrac{30}{5}$

$y = 6$

(d) another version

$\tfrac{1}{2}y + \tfrac{1}{3}y = 5$

$\tfrac{5}{6}y = 5$

$\tfrac{5}{6}y \div \tfrac{5}{6} = 5 \div \tfrac{5}{6}$

$y = 6$

EXERCISE A – 18

In solving an equation, you perform whatever operations are necessary in order to isolate the unknown term on one side (usually, although not necessarily, the left side).

1. Do **not** solve the following equations. Instead, tell what operations are necessary in each case to isolate a on the left side with a single numerical term on the right side.

EXAMPLE

> In the equation $3a - 26 = 49$, the necessary operations are
> (i) Add 26   (ii) Divide by 3.

(a) $a + 11 = 24$

(b) $\dfrac{a}{5} = 12$

(c) $3.5a = 21$

(d) $0.4a = 30$

(e) $a - 16 = 12$

(f) $3a - 23 = 52$

(g) $\dfrac{a}{4} + 7 = 12$

(h) $6a + 9a = 30$

(i) $\frac{2}{3}a = 5.2$

(j) $15a + 12 = 9a - 14$

(k) $\frac{3}{4}a - 14 = 16$

(l) $7a - 16 = 3a$

2. In the following equations the first step is done for you. Complete the solutions.

(a) $m - 15 = 24$
$$m - 15 + 15 = 24 + 15$$

(b) $m + 7 = 11$
$$m + 7 - 7 = 11 - 7$$

(c) $4m = 12$
$$\frac{4m}{4} = \frac{12}{4}$$

(d) $1.2m = 1.8$
$$\frac{1.2m}{1.2} = \frac{1.8}{1.2}$$

(e) $\frac{3}{8}m = 15$
$$\tfrac{3}{8}m \times \tfrac{8}{3} = 15 \times \tfrac{8}{3}$$

(f) $2m - 7 = 8m - 3$
$$2m - 7 + 7 = 8m - 3 + 7$$

3. The roots of the following equations have been worked out as 5. Most are correct; some are incorrect. Substitute 5 as the value of b to see which equations are correct.

EXAMPLE

> The first equation is $4b - 16 = 3b - 11$
> If $b = 5$, the left side is $4(5) - 16$, or $20 - 16$, or 4.
> The right side is $3(5) - 11$, or $15 - 11$, or 4.
> The solution $b = 5$ is correct for this equation.

(a) $4b - 16 = 3b - 11$

(b) $1.8b + 3 = 12$

(c) $0.2b + 3 = 2b - 6$

(d) $\dfrac{b}{10} - 1.5 = 2b - 9$

(e) $\dfrac{b}{2} + 2.5 = b$

(f) $160 = 12b + 100$

(g) $6b - 29 = 2b + 2$ (h) $b + 4 - (2b - 7) = -6(b - 4)$

4. Solve and verify the following equations.

(a) $a + 24 = 49$

(b) $a + 2.5 = 9.25$

(c) $b + 50 = 20$ (Use subtraction)

(d) $k - 16 = 35$

(e) $c - 1.7 = 32$

(f) $c - 8 = {}^-30$ (Use addition)

(g) $5m = 20$

(h) $9\frac{1}{2}n = 47\frac{1}{2}$

(i) $4.2k = 25.2$ (Use division)

(j) $\frac{3}{4}r = 48$

(k) $\frac{5}{12}k = 30$

(l) $\frac{1}{16}p = 4$ (Use multiplication)

(m) $\frac{4}{5}v - \frac{2}{5}v = 28$ (n) $\dfrac{h}{6} = 8$ (o) $3y + 8\frac{1}{2} = 8y - 11\frac{1}{2}$

(p) $0.05a = 2.45$ (q) $b - 1 = {}^-12$ (r) ${}^-24 = {}^-6k$

(s) $m + 0.2m = 60$ (t) $y + \frac{1}{3}y = 8$

(u) $6w - (4w - 4) = {}^-8(w + 2)$ (v) $\dfrac{0.02}{0.8} = \dfrac{n}{10}$

(w) $t - \frac{3}{4}t = 16 - 4$ (x) $2a - 3(a + 5) = 6a - 1$

ANSWERS FOR EXERCISE A - 18

1. (a) Subtract 11   (b) Multiply by 5   (c) Divide by 3.5

(d) Divide by 0.4   (e) Add 16   (f) Add 23 and divide by 3

(g) Subtract 7 and multiply by 4   (h) Combine $6a$ and $9a$. Divide by 15.

(i) Multiply by $\frac{3}{2}$ or divide by $\frac{2}{3}$

(j) Subtract 12; subtract $9a$; divide by 6

(k) Add 14 and multiply by $\frac{4}{3}$ (l) Add 16; subtract $3a$; divide by 4

2. (a) $m = 39$ (b) $m = 4$ (c) $m = 3$ (d) $m = 1.5$

(e) $m = 40$ (f) $m = -\frac{2}{3}$

3. (b) LS: $1.8(5) + 3$ RS: 12   (c) LS: $0.2(5) + 3$ RS: $2(5) - 6$
   $= 9 + 3$ $= 1 + 3$ $= 10 - 6$
   $= 12$ (b) is correct $= 4$ $= 4$ (c) is correct

(d) LS: $\frac{5}{10} - 1.5$ RS: $2(5) - 9$ (e) LS: $\frac{5}{2} + 2.5$ RS: 5
   $= {}^-1$ $= 10 - 9$ $= 5$ (e) is correct
           $= 1$ (d) is incorrect.

(f) correct   (g) incorrect

(h) LS: $(5) + 4 - (2 \times 5 - 7)$ RS: $-6(5 - 4)$
   $= 9 - (10 - 7)$ $= {}^-6(1)$
   $= 9 - 3$ $= {}^-6$ (h) is incorrect.
   $= 6$

4. (a) $a = 25$ (b) $a = 6.75$ (c) $b = {}^-30$ (d) $k = 51$

(e) $c = 33.7$ (f) $c = {}^-22$ (g) $m = 4$ (h) $n = 5$

(i) $k = 6$ (j) $r = 64$ (k) $k = 72$ (l) $p = 64$

(m) $v = 70$ (n) $h = 48$ (o) $y = 4$ (p) $a = 49$

(q) $b = -11$ (r) $k = 4$ (s) $m = 50$ (t) $y = 6$

(u) $w = {}^-2$ (v) $n = 0.25$ (w) $t = 48$ (x) $a = {}^-2$

For more equations see page 222.

ITEM A – 19 REARRANGING A FORMULA

A formula is a special form of equation in that it states a mathematical rule having to do with the relationship between two quantities. One of the most familiar formulas, $A = lw$, defines the relationship between the area of a rectangle and the dimensions of the rectangle. (A stands for area, l and w represent length and width respectively.) In the formula $A = lw$, the area is expressed as the product of the length and the width. But sometimes it is necessary to express the length in terms of area and width, or the width in terms of area and length. This is called changing the subject of a formula, or rearranging a formula. The procedure is the same as that for the solution of any equation; the only difference is that a formula has more literal terms and fewer numerical terms than most other equations.

STEPS IN REARRANGING A FORMULA

One: *Reverse the formula so that the new subject is on the left side.*

Two: *Analyse the way in which the other terms on the left side are combined with the new subject.*

Three: *Use whatever inverse operations are necessary to remove these terms (and, of course, perform the duplicate operation on the other side).*

EXAMPLE

In the formula $A = lw$, solve for w.

STEPS

One: Reverse the formula. $lw = A$

Two: l and w are joined by multiplication.

Three: Divide each side by l. $\dfrac{\cancel{l}w}{\cancel{l}} = \dfrac{A}{l}$

The width of a rectangle is the area divided by its length. $w = \dfrac{A}{l}$

A SLIGHTLY MORE DIFFICULT EXAMPLE

Solve for t in the formula $A = p + prt$.

STEPS

One:	Reverse the formula.	$p + prt = A$
Two:	(a) p and t (in prt) are joined by addition.	$p + prt - p = A - p$ $prt = A - p$
	(b) rp is joined to t by multiplication.	$\dfrac{\cancel{prt}}{\cancel{pr}} = \dfrac{A - p}{pr}$
Three:	(a) subtract p (b) divide by pr	$t = \dfrac{A - p}{pr}$

EXERCISE A -- 19

1. Solve for h in the formula $A = bh$.
2. Solve for r in the formula $i = prt$.
3. Solve for d in the formula $c = \pi d$. (Handle π exactly as you would any other number.)
4. Solve for h in the formula $A = \pi r^2 h$.
5. Solve for v in the formula $a = \dfrac{v}{t}$
6. Solve for w in the formula $p = 2(l + w)$.
7. Solve for h in the formula $A = \dfrac{h}{2}(x + y)$.
8. Solve for C in the formula $F = \frac{9}{5}C + 32$.
9. Solve for c in the formula $p = a + b + c$.

ANSWERS FOR EXERCISE A – 19

1. $A = bh$

   $bh = A$

   $$\frac{\cancel{b}h}{\cancel{b}} = \frac{A}{b}$$

   $$h = \frac{A}{b}$$

2. $i = prt$

   $prt = i$

   $$\frac{\cancel{p}r\cancel{t}}{\cancel{p}\cancel{t}} = \frac{i}{pt}$$

   $$r = \frac{i}{pt}$$

3. $c = \pi d$

   $\pi d = c$

   $$\frac{\pi d}{\pi} = \frac{c}{\pi}$$

   $$d = \frac{\cdot c}{\pi}$$

4. $A = \pi r^2 h$

   $\pi r^2 h = A$

   $$\frac{\pi r^2 h}{\pi r^2} = \frac{A}{\pi r^2}$$

   $$h = \frac{A}{\pi r^2}$$

5. $$a = \frac{v}{t}$$

   $$\frac{v}{t} = a$$

   $$\frac{v}{t} \times t = a \times t$$

   $$v = at$$

6. $p = 2(l + w)$

   $$\frac{2(l + w)}{2} = \frac{p}{2}$$

   $$l + w = \frac{p}{2}$$

   $$w = \frac{p}{2} - l$$

7. $$A = \frac{h}{2}(x + y)$$

   $$\frac{h}{2}(x + y) = A$$

   $$\frac{h(x + y)}{2} \times 2 = A \times 2$$

   $$h(x + y) = 2A$$

   $$\frac{h\cancel{(x + y)}}{\cancel{(x + y)}} = \frac{2A}{(x + y)}$$

   $$h = \frac{2A}{x + y}$$

8. $F = \frac{9}{5}C + 32$

   $\frac{9}{5}C + 32 = F$

   $\frac{9}{5}C + 32 - 32 = F - 32$

   $\frac{9}{5}C = F - 32$

   $\frac{9}{5}C \times \frac{5}{9} = \frac{5}{9}(F - 32)$

   $C = \frac{5}{9}(F - 32)$

9. $p = a + b + c$

   $a + b + c = p$

   $a + b + c - (a + b) = p - (a + b)$

   $c = p - a - b$

POST TEST: PART TWO

MEASUREMENT

M -- 1 Complete the following tables.

(a) 1 mi. = _____ yd.   (b) 1 sq. yd. = _____ sq. in.

(c) 1 T = _____ cwt.

M -- 2 Change the following units of measure.   *seconds*

(a) 3.25 sq. ft. = _____ sq. in.   (b) 3720 s = _____ h

M -- 3 (a) Subtract 1 yd. 2 ft. 11 in. from 6 yd. 1 in.

(b) Multiply 3 qt. 1.5 pt. by 6.

M -- 4 Name the plane figures illustrated below.

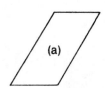   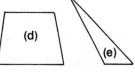

M -- 5 Name the plane figure for which the formula is given.

(a) $p = 2(l + w)$ (b) $A = s^2$ (c) $A = bh$ (d) $A = \dfrac{bh}{2}$

M -- 6 Find the area of

(a) a circle with circumference 69.08 cm

(b) a trapezoid with bases 14" and 16", and height 11".

M -- 7 (a) One angle of an equilateral triangle is 60°. What are the sizes of the other two angles?

(b) Find the area of a right triangle if the hypotenuse is 35 cm, and one side is 21 cm.

M -- 8 (a) What is π?

(b) What is the radius of a circle with diameter 15.5 cm?

M -- 9 Find the volume of each of the following:

(a) a four-inch cube

(b) a cylinder with a diameter of 20 ft. and a height of 13 ft.

M -- 10 (a) How many square inches are there in 1 sq. yd. + $2\frac{1}{2}$ sq. ft.?

(b) How many square feet are there in $2\frac{5}{9}$ sq. yd.?

M -- 11 Under SI metric, in what units will the following be sold?

(a) fuel oil   (b) cheese   (c) peanuts

(d) shampoo   (e) potatoes

M – 12 Write in correct sequence the six metric prefixes, from 0.001 to 1000.

M – 13 Change the following:

(a) 3200 g to kilograms   (b) 300.42 cm to decimetres

(c) 0.9 L to centilitres.

M – 14 Correct the following to conform with acceptable SI format.

(a) The answer is .435   (b) His speed was 105 Km/hr.

(c) The child weighed $8\frac{1}{2}$ kg.

M – 15 Find (a) the area of a 5.2 m square.

(b) the volume of a cylindrical bottle with diameter 8 cm, and height 9 cm.

ALGEBRA

A – 1 Which of the following pairs are like terms?

(a) 2 and 3m (b) 4xyz and 10xyz

(c) k and 11k (d) gh^2 and g^2h

A – 2 Name the literal coefficients in

(a) 3bc (b) ^-2k (c) p

A – 3 Subtract the absolute value of $^-24$ from the absolute value of $^+24$.

A – 4 Add: $^-2 + {}^-500 + {}^+64 + {}^+450 + 60$

A – 5 Subtract.

(a) $^-2$ from $^-5$ (b) $^+7$ from 0   (c) 100 from $^-38$

A – 6 Multiply.

(a) $^-7$ by $^-2$ (b) $0 \times {}^-7$ (c) $^-7$ by $^-1$ by $^-4$

A – 7 Divide.

(a) 18 by $^-3$ (b) $^-18$ by $^-3$ (c) $^+18$ by $^+3$

(d) $^-18$ by $^+3$

A – 8 Write algebraic expressions for

(a) The sum of three consecutive numbers of which $k + 1$ is the second

(b) the number of miles a car travels in w hours at p miles per hour.

A – 9 Add: $a + 4a + 7a + ab + a^2$

A – 10 Remove brackets and simplify.

$3a(a - 5) + 2(a + 14) - 4(a - 2)$

A – 11 Divide.

(a) 40a by 4a (b) 16ab by b^2 (c) 36a^2b^3c by abc^2

A – 12 Find the result if $a = 7$: $a^5 \times a^3 \div \dfrac{a^8}{a^2}$

A – 13 Find the results.

(a) $a + {}^-b + {}^-c$

(b) $a - (b - c)$

(c) $ab \times {}^-c$

(d) ${}^-a \div ({}^-b)({}^+c)$

A – 14 (a) Add: $4(a - b) + (6a + 8b) + 3(2a - 7b)$

(b) Subtract. (i) k from 0   (ii) $k + m$ from 0   (iii) k from $(k + m)$

(iv) ${}^-3k - 4m$ from $2k + 5m$

A – 15 Multiply $(2a^2 + 2a - 7)$ by $(a - 3)$

A – 16 Divide $(4m \times 2mn \times 3m^2n)$ by $(18m^3n \div 2mn^2)$.

A – 17 Find the result of each expression if $r = 4$, $s = {}^-2$, and $t = \frac{1}{2}$.

(a) $3r^2s$ (b) $2st - 4r$ (c) $r^2 + \dfrac{s}{r} - t^2 \times r$

A – 18 Solve for k in each of the following equations.

(a) $k + 12 = 14$ (b) $0.5k + 3.5 = 0.8k - 0.7$ (c) $\dfrac{k}{5} = 13$

(d) $5k - 12 = 2k + 15$ (e) $4k - 2(k - 15) = 7k - 20$

(f) $\frac{7}{9}k = 63$ (g) $15 - 1\frac{1}{4}k = 12 - 4\frac{1}{4}k$

A – 19 Rearrange the formula $A = \dfrac{h}{2}(x + y)$, so that the subject is x.

ANSWERS FOR POST TEST: PART TWO

MEASUREMENT	ALGEBRA

MEASUREMENT

M – 1 (a) 1760   (b) 1296   (c) 20

M – 2 (a) 468   (b) 1.033

M – 3 (a) 4 yd. 2 in.

      (b) 5 gal. 2 qt. 1 pt.

M – 4 (a) parallelogram

      (b) rectangle   (c) circle

      (d) trapezoid   (e) triangle

      (e) square

M – 5 (a) rectangle

      (b) square

      (c) parallelogram (d) triangle

M – 6 (a) 379.94 cm^2

      (b) 165 sq. in.

M – 7 (a) 60°   (b) 294 cm^2

M – 8 (a) $\dfrac{\text{circumference}}{\text{diameter}}$

      (b) 7.75 cm

M – 9 (a) 64 cu. in.

      (b) 4082 cu. ft.

M – 10 (a) 1656   (b) 23

M – 11 (a) litre (b) gram (c) gram

      (d) millilitre   (e) kilogram

M – 12 milli, centi, deci, deca, hecto, kilo.

M – 13 (a) 3.2 kg   (b) 30.042 dm

      (c) 90 cL

M – 14 (a) 0.435   (b) 105 km/h

      (c) 8.5 kg (mass)

M – 15 (a) 27.04 m^2

      (b) 452.16 cm^3

ALGEBRA

A – 1   (b) and (c) are like terms

A – 2   (a) bc (b) k (c) p

A – 3   0

A – 4 $^+72$

A – 5   (a) $^-3$ (b) $^-7$ (c) $^-138$

A – 6   (a) $^+14$ (b) 0   (c) $^-28$

A – 7   (a) $^-6$ (b) $^+6$

      (c) 6   (d) $^-6$

A – 8   (a) $3k + 3$ (b) wp (miles)

A – 9 $a^2 + 12a + ab$

A – 10 $3a^2 - 17a + 36$

A – 11   (a) 10   (b) $\dfrac{16a}{b}$

      (c) $\dfrac{36ab^2}{c}$

A – 12   49

A – 13   (a) $a - b - c$ (b) $a - b + c$

      (c) ^-abc (d) $\dfrac{a}{bc}$

A – 14   (a) $4a - 17b$

      (b) (i) ^-k (ii) $^-k - m$

      (iii) m (iv) $5k + 9m$

A – 15 $2a^3 - 4a^2 - 13a + 21$

A – 16 $\dfrac{8m^2n^3}{3}$

A – 17   (a) $^-96$ (b) $^-18$

      (c) $14\frac{1}{2}$

A – 18   (a) $k = 2$ (b) $k = 14$

      (c) $k = 65$ (d) $k = 9$

      (e) $k = 10$ (f) $k = 81$

      (g) $k = ^-1$

A – 19 $x = \dfrac{2A}{h} - y$

PRACTICE EXERCISES: WHOLE NUMBERS

W – 1 PLACE VALUE

1. Give the place value of each underlined digit.

   (a) 4<u>6</u>02   (b) <u>3</u> 721 245   (c) <u>6</u>24 <u>0</u>08 095 70<u>0</u>

2. (a) Write the number that has 4 in the tens of thousands place, 6 in the units place, 9 in the millions place, and 0 in all other places.

   (b) Write the number composed of six millions, seven tens, four ones, nine thousands, three hundreds of thousands, two hundreds, and five tens of thousands.

W – 2 READING LARGE NUMBERS

1. Copy and read aloud each of the following numbers while a friend checks the answers on next page.

   (a) 3 075 020   (b) 600 000 000   (c) 9 042 000 750

   (d) $82 674 008.24   (e) 7 003 003   (f) 140 000 140

   (g) 2 002 002   (h) 524 000   (i) 200 000 000 000

2. Turn back to Exercise W – 3 on page 10. Cover the words while you read the numbers in the column at the right. Check for accuracy.

W – 3 WRITING LARGE NUMBERS

Write numerals for each of the following:

(a) three thousand, fifteen

(b) two million, three hundred thousand

(c) six billion

(d) eight billion, four hundred

(e) eleven million, eleven thousand, eleven

(f) seventy-two thousand, sixty

(g) nine hundred billion

(h) two hundred forty-seven million, one hundred

(i) six billion, eighty-five thousand

(j) ninety-four thousand, one

Answers for whole number exercises are on pages 223 and 224.

ANSWERS FOR W – 2

(a) three million, seventy-five thousand, twenty   (b) six hundred million

(c) nine billion, forty-two million, seven hundred fifty

(d) eighty-two million, six hundred seventy-four thousand, eight dollars and twenty-four cents

(e) seven million, three thousand, three

(f) one hundred forty million, one hundred forty

(g) two million, two thousand, two   (h) five hundred twenty-four thousand

(i) two hundred billion

W – 5 TERMINOLOGY

1. Describe each number between 50 and 60 as (a) odd or even (b) prime or composite.

2. Give one factor to prove that each of the following is composite.

   (a) 70   (b) 15   (c) 27   (d) 95   (e) 51

3. Why are numbers that end in 0 or 5 composite?

4. Give a common factor for each pair.

   (a) 50 and 70   (b) 65 and 78   (c) 18 and 81

5. Find the average for each set.

   (a) 90, 60, and 30   (b) 8, 3, 6, 5, 0, 2, and 4

   (c) 42, 61, and 38   (d) $2.84, $4.00, and $3.33

6. Find the value of

   (a) $\dfrac{10^2}{5^2}$ (b) $3^5 - 2^6$ (c) $14 + 6^3$ (d) $(14 + 6)^3$ (e) $7^2 \div 6^2$

   (f) $1^3 + 3^1$ (g) $10^5 - 10^4$ (h) $10^5 \div 10^4$

7. Which is larger,

   (a) a power with base 7 and exponent 3, or a power with base 3 and exponent 7?

   (b) 10^5 or 5^{10}?

8. Write the square root of

   (a) 64   (b) 121   (c) 400   (d) 225   (e) 81   (f) 1

9. Calculate each result using the correct order of operations.

   (a) $240 + 4 \times 6$ (b) $16 \div (5 + 3) \times 6$

   (c) $6 \times 6 - 4 \div 2 + 8 \times 6$ (d) $\frac{72}{8} + 160 \div 8$

   (e) $2(4^2 - 9) + 3(8 + 1)^2$ (f) $6^2(8 + 6)$

   (g) $6 \times 2^3 \div 24 + 3(5 + 6 \times 12)$ (h) $3 \times 10^2 + 4 \times 10 + 2$

   (i) $15 + \frac{1}{2}$ of $20 \div 2$ (j) $(6 + 9)(8 - 2)$

   (k) $\sqrt{196} - (2^3 + 5)$

W – 6 PART ONE: ADDITION

1. 8 + 14 + 63 + 72 + 27   2. 95 + 361 + 2048 + 6

3. 756 + 224 + 993   4. 175 + 384 + 69 + 775 + 80

5. 36 124 + 71 + 233 + 12 986 + 7 662

6. Add vertically and horizontally.

    (a)   6 +   75 + 492 + 721 +   57
          26 +   84 +   15 +   8 + 224
           9 + 227 +   35 + 694 +   16
          43 +   6 + 865 +   38 +   83
         800 + 273 +   72 +   9 + 976

    (b) 2389 +   777 +   496 + 6542
         6488 + 9026 + 9982 +   166
           17 +   339 + 6567 +   278
          394 + 1762 +   36 + 4777
         1066 +   73 +   149 +   654

7. Add 987 to

    (a) 4328   (b) 695   (c) 2479   (d) 6898   (e) 99

8. Find the sums.

    (a)   4261   (b)   7527   (c)   7695   (d)   8869   (e)   9585
           3957   9363   8372   2748   7649
           6845   7846   8365   9567   3728
           1272   8284   8470   7777   8648
          +9944   +8765   +9898   +6425   +5555

    (f)   8859   (g)   6592
           2794   9871
           9664   6994
           7899   8755
          +7340   +9001

W – 6 PART TWO: SUBTRACTION

1. Subtract.

(a) 6000   (b) 8506   (c) 7694   (d) 2300
    ⁻1492   ⁻1193   ⁻1999   ⁻1608

(e) 30 000   (f) 7060   (g) 50 910   (h) 63 812
    ⁻18 291   ⁻1285   ⁻16 872   ⁻4 297

(i) 82 406   (j) 4645   (k) 9111   (l) 80 307
    ⁻1 798   ⁻2000   ⁻789   ⁻65 132

2. Calculate the change from a ten-dollar bill for each of the following amounts:

    (a) $2.47   (b) $0.91   (c) 68¢   (d) $9.23   (e) $5.01

3. Complete the charts.

	Minuend	Subtrahend	Difference
(a)	4000	2765	
(b)	6400		2612
(c)		898	999
(d)	3501	1897	
(e)	23 000		15 869

	Minuend	Subtrahend	Difference
(f)	6200	4309	
(g)	5000		1169
(h)		1600	248
(i)	1297	1198	
(j)	4006		1347

W – 6 PART THREE: MULTIPLICATION

	Set One	Set Two	Set Three	Set Four
(a)	240 × 260	585 × 78	795 × 83	20 × 60 × 30
(b)	320 × 70	469 × 72	888 × 77	4080 × 602
(c)	695 × 84	680 × 40	390 × 370	10 006 × 805
(d)	756 × 38	765 × 39	597 × 540	5007 × 604
(e)	629 × 59	400 × 300	956 × 42	6090 × 807
(f)	800 × 600	666 × 44	784 × 85	3002 × 6004
(g)	524 × 65	580 × 520	720 × 30	90 × 100 × 60
(h)	999 × 77	973 × 68	629 × 47	4007 × 1604
(i)	289 × 46	659 × 59	600 × 500	5236 × 2503
(j)	783 × 95	584 × 79	299 × 48	4600 × 801

W – 6 PART FOUR: DIVISION

	Set One	Set Two	Set Three
(a)	36)7200	632)68 094	6842 ÷ 27
(b)	12)2460	583)89 716	5612 ÷ 39
(c)	36)7272	297)95 067	4972 ÷ 78
(d)	15)75 090	834)36 415	61 003 ÷ 85
(e)	24)14 496	567)71 629	44 699 ÷ 63
(f)	27)10 881	621)30 048	290 464 ÷ 58
(g)	55)220 330	830)64 927	662 865 ÷ 73
(h)	16)6432	724)29 169	258 344 ÷ 86
(i)	33)6666	165)83 627	890 208 ÷ 99
(j)	25)20 175	264)91 957	65 026 ÷ 13

ANALYZING PROBLEMS

Read each problem; then tell whether you would add, subtract, multiply, or divide to find the answer. See the answer for question 1 at end of exercise.

1. If you know how much money a person spent and how much he had left, how would you find how much he had in the first place?

2. If you know a man's yearly income, and the amount that he saves in a year, how would you find how much he spends?

3. If you know the number of students registered in each program at a college, how would you find the total number of students at the college?

4. If you know your height in feet, how do you find your height in inches?

5. If you know how far a car travels on one litre of gas, how do you calculate the number of litres required for a 600 km trip?

6. If you know the starting time and the finishing time of a movie, how do you find the length of the movie?

7. If you're saving to buy a car and you know how much money you can afford to set aside each week, how do you determine the time it will take to save enough money?

8. If you know the number of calories in a whole pie, how do you calculate the number of calories in your piece?

9. If you know your weekly salary, how do you find your annual salary?

10. If you know the number of items a dealer sells, and the total profit that he makes on these items, how do you find the profit on each one?

ANSWER FOR QUESTION 1.

1. You would *add* the amount that he spent and the amount that he had left.

WHOLE NUMBER PROBLEMS

1. A grocer pays $6.84 for three dozen cans of soup, and sells them at 26¢ each. What is his profit?

2. Your bank balance on May 31 was $247.61. During the month of June, you deposited $46.28 and $39.65. You also wrote cheques for $9.71 and $52.16. What was your balance at the end of June?

3. Twelve members of The Fitness Club jogged a total distance of 252 miles in 7 days. What was the average number of miles per member per day?

4. Find the total distance in each case.

	Vehicle	Average speed	Number of hours	Total distance
(a)	Bike	7 mph	4	
(b)	Car	51 mph	16	
(c)	Train	84 mph	7	
(d)	Plane	672 mph	9	

5. Thomas Edison was born in 1847. He was 84 years of age when he died. What was the year of his death?

6. An automobile bought for $6540 was sold three years later for $3270. What was the average yearly depreciation?

7. At a monthly wage of $900.00, how long must I work to earn $31 050.00?

8. A litre of ice cream yields 8 servings. At 96¢ a litre, how much will it cost to serve ice cream to 160 people?

9. A Manpower training program lasted for 36 weeks. During that time there were 7 statutory holidays. If a particular trainee was absent for nine days during the program, on how many days did he actually attend classes?

10. A supermarket special advertises soft drinks at 19¢ each or $3.99 for a case of twenty-four. What is the saving per case?

11. A union negotiated a 32¢ hourly wage increase for each of 127 employees in a factory. What was the cost to the company for one year, assuming that each employee worked a $37\frac{1}{2}$ hour week for 50 weeks per year?

12. Three feet of table space is allowed for each person at a banquet. How many people could be seated at both sides of a thirty-foot table?

13. You can buy a monthly pass on the subway for $26.00, or you can pay 50¢ each time you ride. If you use the subway 69 times during a month, how much do you save by buying the pass?

PRACTICE EXERCISES: COMMON FRACTIONS

F – 1   MEANING AND TERMINOLOGY

1. Write a fraction with numerator 7, and denominator 3.
2. Is the fraction in question 1 proper or improper?
3. Identify the proper fractions in the following set:

$\frac{3}{10}$ $\frac{4}{4}$ $\frac{11}{3}$ $\frac{6}{9}$ $\frac{16}{9}$ $\frac{1}{9}$ $\frac{11}{9}$ $\frac{21}{28}$ $\frac{14}{3}$ $\frac{6}{6}$ $\frac{12}{15}$ $\frac{1}{2}$ $\frac{121}{132}$ $\frac{200}{100}$ $\frac{156}{10}$

4. Change the improper fractions in question 3 to whole or to mixed numbers.
5. Reduce all the proper fractions in question 3 to lowest terms.
6. Identify the unit fractions in question 3.
7. Change the following mixed numbers to improper fractions.

$2\frac{1}{10}$ $7\frac{3}{5}$ $9\frac{2}{3}$ $1\frac{11}{16}$ $8\frac{5}{8}$ $7\frac{1}{12}$ $6\frac{9}{10}$ $4\frac{3}{4}$ $8\frac{5}{16}$ $9\frac{11}{40}$ $7\frac{1}{2}$ $16\frac{4}{5}$

F – 2   EQUIVALENT FRACTIONS

1. Complete this sentence: Equivalent fractions have the same _____ but different _____.

2. Change each of the following fractions to an equivalent fraction with a denominator of 72.

   (a) $\frac{5}{8}$ (b) $\frac{1}{4}$ (c) $\frac{11}{12}$ (d) $\frac{19}{24}$
   (e) $\frac{1}{6}$ (f) $\frac{4}{9}$ (g) $\frac{4}{3}$

3. Complete the following so that all fractions are equivalent.

$$\frac{2}{3} = \frac{}{6} = \frac{}{12} = \frac{}{9} = \frac{}{24} = \frac{}{48} = \frac{}{51} = \frac{}{27}$$

4. Which two fractions are equivalent in each set below?

   (a) $\frac{12}{16}, \frac{15}{20}, \frac{18}{28}$ (b) $\frac{27}{30}, \frac{34}{40}, \frac{9}{10}$
   (c) $\frac{63}{72}, \frac{49}{56}, \frac{70}{84}$ (d) $\frac{90}{120}, \frac{135}{150}, \frac{45}{60}$

5. Which fraction in each set is different in value from the others?

   (a) $\frac{18}{27}, \frac{4}{6}, \frac{10}{15}, \frac{15}{25}$ (b) $\frac{80}{100}, \frac{28}{35}, \frac{9}{15}, \frac{48}{60}$ (c) $\frac{100}{120}, \frac{8}{16}, \frac{20}{24}, \frac{15}{18}$

F – 3   ADDITION

1. $11\frac{11}{36} + 9\frac{1}{9} + 12\frac{5}{12} + 2\frac{2}{3} + 4\frac{3}{4}$ 2. $8\frac{1}{10} + 7\frac{7}{30} + 8\frac{5}{6} + 2$
3. $24\frac{5}{16} + 11\frac{3}{4}$ 4. $8\frac{29}{32} + 17\frac{17}{32} + 19\frac{11}{32} + 4\frac{7}{32}$
5. $\frac{3}{5} + \frac{7}{8} + \frac{9}{16}$ 6. $2\frac{2}{9} + 4\frac{3}{5} + 8\frac{2}{15}$
7. $22 + 1\frac{5}{12} + 4$ 8. $4\frac{1}{3} + 11 + 2\frac{2}{5} + \frac{14}{15}$

Answers for fraction exercises are on pages 225 and 226.

9. $9\frac{3}{5} + 3\frac{3}{10}$

10. $146 + \frac{11}{12}$

11. $24 + 4\frac{2}{3} + \frac{7}{8} + 3\frac{7}{12}$

12. $6\frac{1}{5} + 2\frac{2}{3} + 11\frac{4}{15}$

13. $9\frac{1}{6} + 2\frac{2}{3} + 14$

14. $\frac{2}{5} + \frac{2}{5}$

15. $6\frac{7}{9} + 2\frac{2}{3} + 15 + 6\frac{5}{6}$

16. $140 + 2\frac{2}{3}$

17. $\frac{3}{4} + \frac{5}{12} + \frac{4}{5} + \frac{9}{10}$

18. $2\frac{5}{8} + 16 + 1\frac{1}{5} + \frac{3}{4}$

19. $\frac{1}{3} + \frac{1}{3} + \frac{1}{3}$

20. $2\frac{2}{3} + 9\frac{1}{4} + 5\frac{5}{6}$

21. $11 + 6\frac{1}{2} + 5\frac{3}{8}$

22. $\frac{7}{8} + \frac{7}{8}$

23. $\frac{3}{10} + \frac{5}{6} + \frac{1}{3} + \frac{2}{5}$

24. $1\frac{1}{10} + 9\frac{3}{4} + 2 + 14\frac{4}{5} + \frac{1}{8}$

25. $2\frac{5}{6} + 9\frac{1}{3} + 11 + 8\frac{3}{5}$

26. $341 + 8\frac{3}{4}$

27. $\frac{3}{10} + \frac{4}{10} + \frac{5}{10}$

28. $1\frac{1}{3} + 4\frac{5}{12} + 3\frac{5}{6}$

29. $4\frac{1}{4} + 6\frac{7}{8} + 2\frac{2}{3}$

30. $21\frac{2}{3} + 46\frac{3}{5}$

F – 4 SUBTRACTION

Set One

1. Subtract $9\frac{5}{8}$ from 16.

2. Subtract 9 from $16\frac{5}{8}$.

3. $68 - 16\frac{7}{10}$

4. $20\frac{1}{8} - 16\frac{11}{12}$

5. $32\frac{7}{18} - 13\frac{11}{18}$

6. $4\frac{1}{5} - 2\frac{4}{5}$

7. $6\frac{2}{3} - 6\frac{2}{3}$

8. $8 - \frac{3}{5}$

9. $20 - 2\frac{7}{16}$

10. $17\frac{2}{3} - 4\frac{5}{8}$

Set Two

1. Subtract 16 from $20\frac{3}{4}$.

2. Subtract $16\frac{3}{4}$ from 20.

3. $6\frac{1}{3} - 6\frac{5}{15}$

4. $40 - 16\frac{2}{3}$

5. $40\frac{2}{3} - 16$

6. $4\frac{7}{8} - 1\frac{3}{4}$

7. $2\frac{19}{60} - 1\frac{19}{60}$

8. $16\frac{11}{15} - 9\frac{14}{15}$

9. $32\frac{1}{2} - 32\frac{4}{8}$

10. $\frac{7}{8} - \frac{7}{16}$

Set Three

1. $4\frac{7}{16} - 1\frac{3}{16}$

2. $20\frac{3}{4} - 11$

3. $16 - 3\frac{5}{8}$

4. $16\frac{5}{8} - 3$

5. $1\frac{7}{10} - 1\frac{63}{100}$

6. From $12\frac{3}{4}$ subtract the sum of $4\frac{1}{2}$ and $2\frac{5}{8}$.

7. Take 21 from $28\frac{2}{3}$; then subtract $5\frac{5}{6}$ from the answer.

8. If you had walked $11\frac{1}{4}$ miles of a 40-mile hike, how far would you still have to go?

9. $6\frac{1}{4} + 2\frac{2}{5} - 7\frac{1}{5} + 8 - 2\frac{1}{2} - \frac{3}{4}$

10. $25 - 10\frac{2}{3} + 9\frac{1}{4} - 11\frac{11}{12} + \frac{5}{12} - 12$

F – 5 MULTIPLICATION

Set One: Cancellation Practice

1. $\dfrac{24 \times 8 \times 15}{18 \times 5 \times 12}$

2. $\dfrac{24}{25} \times \dfrac{15}{16} \times \dfrac{7}{12}$

3. $\dfrac{52 \times 17 \times 28}{14 \times 39 \times 51}$

4. $\dfrac{54}{9} \times \dfrac{36}{9} \times \dfrac{21}{7}$

5. $\dfrac{60 \times 18 \times 12 \times 16}{144 \times 30}$

6. $\dfrac{25}{30} \times \dfrac{36}{35} \times \dfrac{63}{27}$

7. $\dfrac{27 \times 64 \times 72 \times 87}{24 \times 58 \times 81 \times 48}$

8. $\dfrac{100 \times 200 \times 40 \times 60}{3200 \times 50 \times 80}$

9. $\dfrac{14 \times 6 \times 9 \times 27}{18 \times 3 \times 63}$

10. $\dfrac{78 \times 21}{2}$

Set Two

1. $\frac{1}{9} \times 4\frac{2}{3} \times 5\frac{1}{7} \times \frac{3}{4}$

2. $7 \times 8 \times \frac{7}{8}$

3. $9 \times \frac{17}{18}$

4. $400 \times 6\frac{3}{40}$

5. $3006 \times 2\frac{1}{6}$

6. $\frac{1}{3} \times \frac{4}{10} \times \frac{15}{24} \times \frac{3}{4}$

7. $3\frac{3}{8} \times 4\frac{3}{10} \times \frac{5}{9}$

8. $2\frac{5}{8} \times \frac{4}{7} \times 2\frac{2}{5} \times 1\frac{1}{6}$

9. $\frac{5}{8}$ of 144

10. $2\frac{1}{2} \times 2\frac{1}{2}$

Set Three

1. $1600 \times 3\frac{3}{4}$

2. $\frac{11}{12}$ of 72

3. $\frac{1}{2} \times \frac{1}{2} \times \frac{1}{2}$

4. $3\frac{1}{5} \times 2\frac{7}{8} \times \frac{9}{10}$

5. $\frac{11}{16} \times 64$

6. $2\frac{2}{9} \times \frac{18}{25} \times 1\frac{7}{8}$

7. $5\frac{3}{5} \times 3\frac{4}{7} \times \frac{4}{15}$

8. $5\frac{1}{3} \times 6\frac{3}{10} \times 7\frac{1}{2}$

9. $\frac{3}{4}$ of $36 \times 3\frac{1}{3}$

10. $365 \times 4\frac{1}{5}$

Set Four

1. Subtract the product of $\frac{4}{9}$ and $\frac{3}{4}$ from the sum of $\frac{5}{6}$ and $\frac{4}{9}$.

2. Add the difference between $8\frac{1}{4}$ and $2\frac{2}{3}$ to the product of $1\frac{1}{6}$ and $1\frac{3}{4}$.

3. Seven-eighths of the students in a machine shop program are men. If the total number of students is 32, how many are women?

4. If you save $\frac{2}{25}$ of your monthly salary of $1250, how much would you save in one year?

5. How far does a plane travel in 4 h 20 min at an average speed of 540 miles per hour?

6. Ken put $\frac{1}{4}$ of a melon on a plate. Bill took one-half of it. What fraction of the whole melon did Bill take?

7. A metal bar weighs $3\frac{1}{4}$ lb. per foot. What is the weight of a four-foot bar?

F – 6 DIVISION

Set One

1. $34 \div 8\frac{1}{2}$ 2. $28 \div 2\frac{2}{3}$ 3. $2\frac{7}{10} \div 3$

4. $\frac{1}{2} \div \frac{1}{6}$ 5. $7 \div 1\frac{2}{5}$ 6. $6\frac{3}{5} \div 2\frac{1}{5}$

7. $24 \div \frac{2}{3}$ 8. $\frac{3}{4} \div \frac{3}{8}$ 9. $\frac{1}{24} \div \frac{1}{3}$

Set Two

1. $105 \div 1\frac{1}{6}$ 2. $65 \div \frac{1}{5}$ 3. $4\frac{5}{6} \div 8\frac{2}{7}$

4. $5 \div \frac{5}{8}$ 5. $\frac{15}{22} \div \frac{9}{11}$ 6. $2\frac{1}{5} \div \frac{3}{10}$

7. $\frac{2}{3} \div 24$ 8. $15 \div \frac{3}{5}$ 9. $36 \div 1\frac{1}{2}$

Set Three

1. $\frac{1}{8} \div 3$ 2. $\frac{4}{5} \div \frac{5}{8}$ 3. $9 \div 3\frac{1}{5}$

4. $14\frac{2}{3} \div 11$ 5. $3\frac{1}{4} \div 3\frac{1}{4}$ 6. $27 \div \frac{9}{10}$

7. $4\frac{1}{8} \div 1$ 8. $2\frac{1}{8} \div 2\frac{5}{6}$ 9. $8 \div 1\frac{3}{5}$

PROBLEMS WITH FRACTIONS

1. A builder buys two pieces of land at \$4500 per acre. If the pieces are $95\frac{1}{2}$ acres and $105\frac{3}{4}$ acres, what does he pay for both?

2. John is 5 ft. $2\frac{1}{2}$ in. tall. Bob is 6 ft. How much taller is Bob?

3. How many tape labels, each $1\frac{3}{4}$ in., can be cut from a three-foot strip of plastic?

4. How many guests can be served from 24 lb. of steak if each serving is three eighths of a pound?

5. Plastic containers each hold $1\frac{5}{8}$ cups of juice. How much juice is needed to fill 28 containers?

6. Find the average weight of four infants whose weights are $6\frac{1}{2}$ lb., 7 lb. 4 oz., $7\frac{3}{4}$ lb., and 8 lb.

7. A recipe which calls for $1\frac{1}{2}$ cups of sugar makes 4 dozen cookies. How much sugar would be needed to make 12 dozen cookies?

8. Three people paid $\frac{3}{10}$ of their salaries for rent. What monthly rent did each pay at the following salaries?

   (a) \$1054   (b) \$900   (c) \$1270

9. A wooden pole, $10\frac{1}{2}$ ft. long, is cut into six equal pieces. How long is each piece?

10. When preserving berries $2\frac{1}{2}$ cups of sugar are required for each quart. How much sugar is required for $6\frac{1}{4}$ qt. of berries?

11. A women's club bought 63 lbs. of chocolate to be re-sold in packages of $4\frac{1}{2}$ oz. How many packages would they get from 63 lbs.?

PRACTICE EXERCISES: DECIMALS

D – 1 READING DECIMAL NUMBERS

Cover the answers below while you read each number. Check for accuracy.

(a) 0.023	(b) 6.0004	(c) 0.7	(d) 2.05
(e) 0.008 10	(f) 9.999	(g) 100.100	(h) 10.853
(i) 620.0014	(j) 0.000 008	(k) 2357.8	

ANSWERS

(a) twenty-three thousandths   (b) six and four ten-thousandths

(c) seven tenths   (d) two and five hundredths

(e) eight hundred ten hundred-thousandths

(f) nine and nine hundred ninety-nine thousandths

(g) one hundred and one hundred thousandths

(h) ten and eight hundred fifty-three thousandths

(i) six hundred twenty and fourteen ten-thousandths

(j) eight millionths

(k) two thousand three hundred fifty-seven and eight tenths

D – 2 WRITING DECIMAL NUMBERS

Cover the answers at the right while you write numerals for each.

(a) forty-two ten-thousandths	0.0042
(b) sixty-five and eleven hundredths	65.11
(c) nine hundred five ten-thousandths	0.0905
(d) nine hundred and five ten-thousandths	900.0005
(e) three thousand and two hundred twenty ten-thousandths	3000.0220
(f) eighty-five millionths	0.000 085
(g) one million, two hundred and six hundred twenty thou- sandths	1 000 200.620
(h) ninety-eight hundred	9800
(i) ninety-eight hundredths	0.98
(j) nine thousand, eight hundred	9800
(k) nine thousand and eight hundredths	9000.08

Answers for decimal exercises are on pages 226—228.

D – 3 CHANGING DECIMAL FRACTIONS TO COMMON FRACTIONS

1. Write equivalent common fractions in lowest terms.

   (a) 0.2   (b) 1.6   (c) 2.08   (d) 16.0025
   (e) 0.125   (f) 0.170   (g) 5.875   (h) 5.0875
   (i) 0.888   (j) 95.95   (k) 0.000 455   (l) 200.075

2. Which numbers in each set below are equal in value?

   (a) 0.2,   0.200, $\frac{1}{5}$ (b) 0.214,   0.2140,   0.0214
   (c) $\frac{3}{4}$,   0.075, $\frac{15}{20}$ (d) $\frac{16}{100}$,   0.016, $\frac{160}{1000}$ (e) $\frac{150}{100}$, $1\frac{1}{2}$,   1.05
   (f) 2.2,   2.22,   2.220

D – 4 ADDITION AND SUBTRACTION

1. $0.4 + 2.16 + 700 - 8.314 - 10 + 0.102$

2. 23 − twenty-eight hundredths   3. nine hundredths $+ 0.05 + 0.86$

4. Subtract seven and eight tenths from 17.

5. Find the sum of 4.86, 27, 3.004, and 2.6.

6. Subtract 4.04 from 4.4.

7. Write the decimal fraction that indicates the size of the shaded portion in each diagram below.

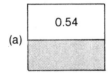 (a)   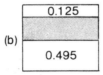 (b)   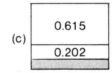 (c)   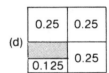 (d)

8. Subtract the sum of 0.4 and 1.26 from the sum of 2.07 and 7.

9. $0.0150 - 0.002$ 10. Add 9.3, 4.031, 0.0126, 72, and 15.5.

11. Subtract 3.75 from $22\frac{1}{2}$.   12. Subtract 0.999 from 10.

13. Add vertically and horizontally; final answer must match.

(a)

1.4 +	22 +	0.65 +	2.7	=
0.06 +	8 +	22.47 +	0.016	=
28 +	0.44 +	50 +	6.375	=
+	+	+		=

(b)

36	+	0.15	+	22.9	+	1.475	=
3.6	+	9.35	+	39	+	0.7	=
0.86	+	120	+	4.070	+	116	=
	+		+		+		=

D – 5 MULTIPLICATION

Set One

1. $0.1 \times 0.6 \times 0.4$
2. 240×1.2
3. 100×5.6
4. 0.125×1.2
5. 800×2.4
6. $400 \times 0.2 \times 0.001$
7. $12 \times 5 \times 0.2 \times 10$
8. 7.652×1000
9. 0.2^3
10. 2.5^2
11. $0.1 \times 0.2 \times 0.3$
12. $0.2^2 \times 10^2$

Set Two

1. 0.045×2.2
2. 31.5×0.011
3. 0.652×10^3
4. 4.5×600
5. $30 \times 0.003 \times 100$
6. $0.06 \times 0.2 \times 0.001$
7. 0.3^4
8. $12^2 + 1.2^2$
9. $450 \times 0.6 \times 0.002$
10. 0.006×0.042
11. $10^2 \times 0.1^2$
12. 51.46×0.25

D – 6 DIVISION

1. In each case, write the new dividend to match the revised divisor.

   (a) $1.6\overline{)40} = 16\overline{)}$
   (b) $2.25\overline{)22.5} = 225\overline{)}$
   (c) $0.08\overline{)16} = 8\overline{)}$
   (d) $0.125\overline{)250} = 125\overline{)}$
   (e) $1.44\overline{)0.72} = 144\overline{)}$
   (f) $0.9\overline{)90} = 9\overline{)}$

2. Divide, then add the quotients in each set.

	Set A	Set B	Set C
(a)	$80 \div 16$	$27 \div 54$	$84 \div 21$
(b)	$800 \div 160$	$2.7 \div 5.4$	$8.4 \div 0.021$
(c)	$0.8 \div 1.6$	$0.027 \div 5.4$	$0.084 \div 210$
(d)	$800 \div 0.016$	$270 \div 5400$	$840 \div 2.1$
(e)	$0.8 \div 1600$	$0.27 \div 0.054$	$0.084 \div 0.021$

3. Divide correct to two places.

(a) $26 \div 34$ (b) $13 \div 1.7$ (c) $0.4 \div 0.37$ (d) $168 \div 2.7$

(e) $3.2 \div 45$ (f) $0.256 \div 71$ (g) $83.2 \div 14.7$ (h) $156 \div 9.4$

4. Write the quotients for Question 1 in this exercise.

D – 7 MULTIPLICATION AND DIVISION

1. Use short cut multiplication or division to write answers for these:

(a) 3.6×10^5 (b) $0.19 \div 1000$ (c) $6250 \div 10^4$

(d) $3.475 \times 1\,000\,000$ (e) $62\,000 \div 10^5$ (f) 0.256×10^2

(g) $4.3 \times 10^4 \div 1000$ (h) $4623 \div 10^3$ (i) $\dfrac{240}{0.24} \times \dfrac{10^2}{10^5}$

2. To change 51 000 to 51, you would divide by 10^3. How would you change each of the following to 51?

(a) 0.0051   (b) 5.1   (c) 0.051   (d) 0.51   (e) 510 000

D – 8 ROUNDING

1. Round to two places.

(a) 1.256   (b) 0.333   (c) 146.248   (d) 0.452 71

2. Round to nearest whole number.

(a) 14.237   (b) 651.8   (c) 1.555   (d) 32 156.74

3. Round to nearest thousandth.

(a) 2.0146   (b) 0.028 15   (c) 4.2   (d) 1600.2513

4. Change to a three-place decimal.

(a) $\frac{1}{12}$ (b) $\frac{7}{32}$ (c) $\frac{7}{9}$ (d) $\frac{11}{15}$

D – 9 CHANGING FRACTIONS TO DECIMALS

1. Write the decimal equivalent for each.

(a) $\frac{3}{4}$ (b) $\frac{1}{10}$ (c) $\frac{3}{5}$ (d) $\frac{9}{25}$ (e) $5\frac{11}{16}$ (f) $3\frac{27}{50}$

2. Write two-place decimal equivalents.

(a) $\frac{2}{3}$ (b) $\frac{5}{7}$ (c) $\frac{13}{15}$ (d) $6\frac{4}{9}$ (e) $11\frac{5}{18}$

3. Write decimal equivalents correct to three places.

(a) $\frac{1}{3}$ (b) $\frac{9}{35}$ (c) $\frac{15}{32}$ (d) $12\frac{5}{6}$ (e) $14\frac{11}{12}$

PRACTICE WITH DECIMALS (I)

1. Write in words.
   (a) 40.0017   (b) 0.002 65
2. Write numerals for: (a) seventeen and seven thousandths
   (b) eight hundred five thousandths   (c) eight hundred and five thousandths
3. Write equivalents using common fractions; reduce if possible.
   (a) 3.24   (b) 1.016   (c) 0.0625   (d) 5.475   (e) 0.004 16   (f) 1.56
4. Calculate the results.
   (a) $200 + 1.4 + 0.0025 + 72.38$ (b) $62 - 0.716$
   (c) $0.1 \times 620 \times 1.2$ (d) $42 \div 0.21$ (e) $0.42 \div 210$
   (f) $\frac{3}{5} \div 0.6$ (g) $1.2 + 0.8 \times 2$ (h) $(1.2 + 0.8) \times 2$
   (i) twenty-seven hundredths $\times$ 0.1   (j) $5\frac{3}{8} \times 100 + 65.7$
5. Change to decimal equivalents.
   (a) $2\frac{9}{10}$ (b) $1\frac{5}{12}$ (c) $6\frac{3}{15}$ (d) $9\frac{7}{40}$ (e) $\frac{5}{7}$ (f) $\frac{19}{32}$
6. Round as indicated.
   (a) 0.157 to nearest hundredth   (b) 1.62 to nearest tenth
   (c) 5.064 38 to three places   (d) 169.9 to the nearest whole number
7. Find the products.
   (a) 1.2×0.006 (b) 12×0.6 (c) 0.0012×600 (d) 0.12×0.6
8. Find the sum of the quotients.
   (a) $96 \div 8$ (b) $0.96 \div 80$ (c) $9600 \div 0.8$ (d) $9.6 \div 0.008$
   (e) $0.0096 \div 80$
9. Find the result. $\dfrac{12.8 \div 4}{8} \div \dfrac{2}{5}$ of $20 + 0.125 \times 3$
10. Solve using decimals.
    (a) $5\frac{3}{4} + 2.7 + 9.03$ (b) $\frac{3}{10}$ of 1.242   (c) $0.45 \div \frac{1}{10}$ (d) $250 \div 1\frac{1}{4}$
11. At 7:00 a.m. the temperature was 15.9°C, but by 1:00 p.m. if had risen to 20.1°C. What was the average hourly temperature change?
12. $\dfrac{0.0625}{0.625} \div \dfrac{1}{0.1}$

PRACTICE WITH DECIMALS (II)

1. Write in words.
   (a) 2.0156   (b) 3000.045
2. Write numerals for
   (a) seven and seven thousandths
   (b) nine thousand and six ten-thousandths
   (c) nine thousand, six, and ten thousandths

3. Write equivalents using common fractions in lowest terms.

   (a) 4.025   (b) 62.083   (c) 0.031 25   (d) 500.08

4. Add and subtract as indicated.

   (a) 0.2 + 0.4 + 0.4   (b) 20 − 1.99   (c) 3.2 + 1.06 − 0.98

5. Multiply.

   (a) 0.3×0.3^2 (b) 1.056 × 0.212   (c) 3.4175×10^5

   (d) six hundredths by one and one tenth   (e) $1\frac{1}{5} \times 2.56$

6. Divide.

   (a) 0.9 ÷ 15   (b) 900 ÷ 0.15   (c) 90 ÷ 1500   (d) 0.009 ÷ 0.15

   (e) $\frac{9}{150}$

7. Round as indicated.

   (a) 14.095 to nearest whole number   (b) 6.38 to nearest tenth

   (c) 0.0158 to two places   (d) 1000.4141 to three places

8. Perform the operations indicated.

   (a) Add $\frac{4}{5}$ and 1.2.   (b) Subtract 7.3 from $9\frac{7}{8}$.   (c) $\frac{3}{4} \times 0.16$.

   (d) Divide 4.4 by one and one tenth.

   (e) Find the average of $3\frac{1}{4}$, 3.125, $2\frac{9}{10}$, and 3.001.

PROBLEMS INVOLVING DECIMALS

1. A gasoline tank has a capacity of 90 L. At the last filling, it took 73.4 L. How much gas was in the tank before it was filled?

2. We drove to Toronto at an average speed of 75.2 km/h. If the trip took 7 h, how far did we drive?

3. It rained six times during July, with recorded rainfalls of 1.4 cm, 1.3 cm, 3.1 cm, 1 cm, 1.6 cm and 0.9 cm. What was the average daily rainfall for the month?

4. The cost of fuel increased from 25.3¢ per litre to 28.1¢ per litre. Find the total increase in cost for a 95-litre tankful of fuel.

5. If 0.24 m of metal costs $12.00, how much does 3 m cost?

6. If 0.2 of a number is 20, what is the number?

7. 0.935 is composed of ___ tenths, ___ hundredths, and ___ thousandths.

8. Arrange the following in order of size from largest to smallest.

   6.5   0.065   650   16   6.6   5.6   6.56

9. If I eat 0.2 of a pie, and you eat 0.125 of it, how much is left?

10. My salary was $1600 a month. If 0.18 of the total was deducted for taxes, 0.63 of the total was spent, and the rest saved, how much money did I save each month?

PRACTICE WITH PERCENT (I)

1. Complete the following charts to show equivalent fractions, decimals, and percents.

	Fraction	Decimal	Percent
(a)	$\frac{2}{5}$	0.4	
(b)		0.08$\dot{3}$	$8\frac{1}{3}\%$
(c)	$\frac{1}{40}$		$2\frac{1}{2}\%$
(d)		0.125	$12\frac{1}{2}\%$
(e)	$\frac{1}{6}$	0.167	

	Fraction	Decimal	Percent
(f)	$\frac{7}{8}$		
(g)		0.175	
(h)			$3\frac{1}{2}\%$
(i)		0.0125	
(j)	$\frac{1}{200}$		

2. Find the total value of each set.

Set A	Set B	Set C
3% of 1500 =	250% of 120 =	0.1% of 8000 =
$2\frac{1}{2}\%$ of 1200 =	$\frac{3}{4}\%$ of 1000 =	$9\frac{1}{4}\%$ of 6800 =
8% of 300 =	0.02% of 4000 =	325% of 760 =
$\frac{1}{2}\%$ of 160 =	$6\frac{1}{2}\%$ of 200 =	1.8% of 13 000 =
300% of 16 = ____	38% of 750 = ____	60% of 324 = ____

3. Six people donated a certain portion of their monthly salaries to a charitable fund. Using the information below, calculate the percent of salary that each person donated.

   A earned $624 and donated $30   B earned $2000 and donated $40
   C earned $850 and donated $36   D earned $1225 and donated $75
   E earned $1200 and donated $100   F earned $1460 and donated $146

4. Each of the items below was reduced by $33\frac{1}{3}\%$. What was the reduced price in each case?

   (a) chair – $200.00   (b) hat – $15.00   (c) chocolates – $8.00
   (d) TV set – $495.00   (e) lampshade – $4.50

5. Complete the following chart; work for (a) and (b) is shown.

	Percent	Partial Amount	Whole Amount (100%)	
(a)	15%		200	15% of 200 = 30
(b)		40	50	$\frac{40}{50} \times 100\% = 80\%$
(c)	10%	6		
(d)	30%	21		
(e)	$6\frac{1}{4}\%$		400	
(f)		$2\frac{1}{2}$	$12\frac{1}{2}$	
(g)	60%		2.4	

Answers for percent exercises are on page 228.

PRACTICE WITH PERCENT (II)

1. Complete the following charts of equivalence.

	Fraction	Decimal	Percent
(a)	$\frac{1}{5}$	0.2	
(b)		0.16$\dot{6}$	$16\frac{2}{3}\%$
(c)	$\frac{7}{16}$		$43\frac{3}{4}\%$
(d)		0.875	$87\frac{1}{2}\%$
(e)	$\frac{5}{12}$	0.417	

	Fraction	Decimal	Percent
(f)	$\frac{4}{25}$		
(g)		0.95	
(h)			$7\frac{3}{4}\%$
(i)		0.0025	
(j)	$\frac{1}{500}$		

2. Calculate the total for each set.

Set A	*Set B*	*Set C*
6% of 8000	9% of 120	1000% of 126
$1\frac{1}{4}\%$ of 2400	$3\frac{3}{4}\%$ of 1600	0.5% of 200
0.2% of 800	0.025% of 20 000	$62\frac{1}{2}\%$ of 40
600% of 25	450% of 100	85% of 700
$83\frac{1}{3}\%$ of 1200	10.2% of 360	1.6% of 3200

3. The selling prices of several articles were reduced 20%. What were the original selling prices if the reduced prices were as follows?

   (a) picture – $160.00   (b) drapes – $240.00   (c) coffee pot – $19.00

   (d) saw – $48.00   (e) rose bush – $4.80   (f) candle – $0.72

4. Complete the following chart.

	Percent	Partial Amount	Whole Amount
(a)	40%		6000
(b)		500	700
(c)	13%	6.5	
(d)		0.24	1.2
(e)	$2\frac{1}{2}\%$		76
(f)	10%	1	
(g)		1200	1500

5. Four people pledged 7% of one week's earnings to the Terry Fox cancer fund. What was the total amount of the pledge if the weekly salaries were as follows?

   (a) $240   (b) $372   (c) $216   (d) $420

PRACTICE EXERCISE: AREA OF PLANE FIGURES

Complete the chart below; write the formula for finding area, then calculate the area.

	Plane Figure	Dimensions	Formula for Area	Area
(a)	Square	side is 0.4 m		
(b)	Rectangle	length is 2.4 cm width is 16 mm		
(c)	Circle	diameter is 2 ft.		
(d)	Triangle	base is 12 in. height is $9\frac{1}{2}$ in.		
(e)	Circle	circumference is 44 in.		(use $\frac{22}{7}$)
(f)	Rectangle	length is 0.2 m width is 0.1 m		
(g)	Trapezoid	bases are 4 in. and 6 in. height is 8 in.		
(h)	Triangle	base is $2\frac{1}{2}$ ft. height is 18 in.		
(i)	Circle	radius is 6 cm		
(j)	Square	side is 1.1 m		
(k)	Parallelogram	base is $5\frac{1}{4}$ in. height is 8 in.		
(l)	Trapezoid	bases are 2.7 cm & 4.3 cm height is 3 cm		

2. For each of the diagrams below, (a) name the plane figure,
   (b) give the formula for finding its area, and   (c) calculate the area.

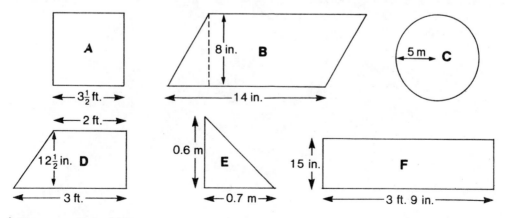

Answers on page 229.

PRACTICE EXERCISE: SOLVING EQUATIONS

Set One
1. $y + 7 = 12$
2. $y - 6 = 42$
3. $y + 4\frac{1}{2} = 16$
4. $y + 2.4 = 8.7$
5. $y - \frac{5}{8} = \frac{7}{16}$

Set Two
1. $3k = 24$
2. $4\frac{1}{2}k = 31.5$
3. $0.6k = 1.2$
4. $68k = 6.8$
5. $2\frac{1}{4}k = 11.25$

Set Three
1. $\dfrac{a}{3} = 32$
2. $\frac{5}{8}a = 2$
3. $\frac{1}{10}a = 1.5$
4. $\dfrac{6a}{5} = 3\frac{3}{5}$
5. $\frac{2}{5}a = 14$

Set Four
1. $3m + 12 = 33$
2. $7m - 4 = 59$
3. $\frac{1}{2}m + 13 = 23$
4. $2.1m - 1.6 = 2.6$
5. $5m + 8 = 23$

Set Five
1. $3n - \frac{1}{2}(n + 6) = n + 4\frac{1}{2}$
2. $4(2n - 5) - 2(n - 3) = 22$
3. $\dfrac{n}{4} + \dfrac{n}{8} = \frac{1}{2}(n - 3)$
4. $3n - 0.2(n + 3) = 2.2$
5. $\frac{3}{4}n + 30 = \frac{5}{8}n + 33$

Set Six
1. $a + b - c = k$ Solve for c.
2. $\dfrac{a}{b} = \dfrac{k}{c}$ Solve for k.

3. $ab = ck$ Solve for b.
4. $ab - c = k$ Solve for c.

5. $abc = k$ Solve for a.
6. $\dfrac{a + b}{c} = k$ Solve for b.

7. $a(b + c) = k - c$ Solve for a.
8. $\dfrac{ab - c}{a} = k$ Solve for b.

9. $bc + a - k = 0$ Solve for a.
10. $bc + a - k = 0$ Solve for c.

Answers on page 229.

ANSWERS FOR SUPPLEMENTARY EXERCISES

WHOLE NUMBERS

W – 1 PLACE VALUE

1. (a) hundreds   (b) 3 millions, 2 tens of thousands
   (c) 6 hundreds of billions, 0 hundreds of millions, 0 units
2. (a) 9 040 006   (b) 6 359 274

W – 3 WRITING LARGE NUMBERS

(a) 3 015   (b) 2 300 000   (c) 6 000 000 000   (d) 8 000 000 400
(e) 11 011 011   (f) 72 060   (g) 900 000 000 000   (h) 247 000 100
(i) 6 000 085 000 (j) 94 001

W – 5 TERMINOLOGY

1. (a) *Odd* are 51, 53, 55, 57, and 59; *even* are 52, 54, 56, and 58.
   (b) *Prime* are 53 and 59; the others are composite.
2. (a) 2 or 5 or 7 or 10 or 14 or 35   (b) 3 or 5   (c) 3 or 9   (d) 5 or 19
   (e) 3 or 17
3. All numbers that end in 0 or 5 are divisible by 5.
4. (a) 2 or 5 or 10   (b) 13   (c) 3 or 9
5. (a) 60   (b) 4   (c) 47   (d) $3.39
6. (a) 4   (b) 179   (c) 230   (d) 8000   (e) $1\frac{13}{36}$ (f) 4
   (g) 90 000   (h) 10
7. (a) 3^7 is larger   (b) 5^{10} is larger
8. (a) 8   (b) 11   (c) 20   (d) 15   (e) 9   (f) 1
9. (a) 264   (b) 12   (c) 82   (d) 29   (e) 257   (f) 504
   (g) 233   (h) 342   (i) 20   (j) 54   (k) 1

W – 6 ADDITION

1. 184   2. 2510   3. 1973   4. 1483   5. 57 076

6. (a) Vertical: 884   665   1479   1470   1356
      Horizontal: 1351   357   981   1035   2130   Total: 5854

   (b) Vertical: 10 354   11 977   17 230   12 417
      Horizontal: 10 204   25 662   7201   6969   1942   Total: 51 978

7. (a) 5315   (b) 1682   (c) 3466   (d) 7885   (e) 1086
8. (a) 26 279   (b) 41 785   (c) 42 800   (d) 35 386   (e) 35 165
   (f) 36 556   (g) 41 213

W – 6 PART TWO: SUBTRACTION

1. (a) 4508   (b) 7313   (c) 5695   (d) 692   (e) 11 709   (f) 5775
   (g) 34 038   (h) 59 515   (i) 80 608   (j) 2645   (k) 8322   (l) 15 175
2. (a) $7.53   (b) $9.09   (c) $9.32   (d) 77¢   (e) $4.99
3. (a) 1235   (b) 3788   (c) 1897   (d) 1604   (e) 7131
   (f) 1891   (g) 3831   (h) 1848   (i) 99   (j) 2659

W – 6 PART THREE: MULTIPLICATION

	Set One	*Set Two*	*Set Three*	*Set Four*
(a)	62 400	45 630	65 985	36 000
(b)	22 400	33 768	68 376	2 456 160
(c)	58 380	27 200	144 300	8 054 830
(d)	28 728	29 835	322 380	3 024 228
(e)	37 111	120 000	40 152	4 914 630
(f)	480 000	29 304	66 640	18 024 008
(g)	34 060	301 600	21 600	540 000
(h)	76 923	66 164	29 563	6 427 228
(i)	13 294	38 881	300 000	13 105 708
(j)	74 385	46 136	14 352	3 684 600

W – 6 PART FOUR: DIVISION

	Set One	*Set Two*	*Set Three*
(a)	200	107 R 470	253 R 11
(b)	205	153 R 517	143 R 35
(c)	202	320 R 27	63 R 58
(d)	5006	43 R 553	717 R 58
(e)	604	126 R 187	709 R 32
(f)	403	48 R 240	5008
(g)	4006	78 R 187	9080 R 25
(h)	402	40 R 209	3004
(i)	202	506 R 137	8992
(j)	807	348 R 85	5002

ANALYZING PROBLEMS

2. Subtract the amount that he saves from the amount of his income.
3. Add the number of students in each program.
4. Multiply your height in feet by 12.
5. Divide 600 km by the number of kilometres the car will travel on one litre.
6. Subtract the starting time from the finishing time.
7. Divide the total amount needed by the amount you can set aside each week.
8. Divide the number of calories by the number of (equal) pieces.
9. Multiply your weekly salary by 52.
10. Divide the total profit by the number of items.

WHOLE NUMBER PROBLEMS

1. $2.52
2. $271.67
3. 3
4. (a) 28 mi. (b) 816 mi. (c) 588 mi. (d) 6048 mi.
5. 1931
6. $1090
7. $34\frac{1}{2}$ mo.
8. $19.20
9. 164
10. $0.57
11. $76 200.00
12. 20
13. $8.50

ANSWERS FOR FRACTION EXERCISES

F - 1 MEANING AND TERMINOLOGY

1. $\frac{7}{3}$ 2. improper   3. $\frac{3}{10}$ $\frac{6}{9}$ $\frac{1}{9}$ $\frac{21}{28}$ $\frac{12}{15}$ $\frac{1}{2}$ $\frac{121}{132}$

4. $\frac{4}{4} = 1$, $\frac{11}{3} = 3\frac{2}{3}$, $\frac{16}{8} = 2$, $\frac{11}{9} = 1\frac{2}{9}$, $\frac{14}{3} = 4\frac{2}{3}$, $\frac{6}{6} = 1$, $\frac{200}{100} = 2$, $\frac{156}{10} = 15\frac{6}{10}$

5. $\frac{6}{9} = \frac{2}{3}$, $\frac{21}{28} = \frac{3}{4}$, $\frac{12}{15} = \frac{4}{5}$, $\frac{121}{132} = \frac{11}{12}$ 6. $\frac{1}{9}$, $\frac{1}{2}$

7. $\frac{21}{10}$, $\frac{38}{5}$, $\frac{29}{3}$, $\frac{27}{16}$, $\frac{69}{8}$, $\frac{85}{12}$, $\frac{69}{10}$, $\frac{19}{4}$, $\frac{133}{16}$, $\frac{371}{40}$, $\frac{15}{2}$, $\frac{84}{5}$

F - 2 EQUIVALENT FRACTIONS

1. value, terms   2. (a) $\frac{45}{72}$ (b) $\frac{18}{72}$ (c) $\frac{66}{72}$ (d) $\frac{57}{72}$ (e) $\frac{12}{72}$ (f) $\frac{32}{72}$ (g) $\frac{96}{72}$

3. $\frac{2}{3} = \frac{4}{6} = \frac{8}{12} = \frac{6}{9} = \frac{16}{24} = \frac{32}{48} = \frac{34}{51} = \frac{18}{27}$

4. (a) $\frac{12}{16}$ and $\frac{15}{20}$ (b) $\frac{27}{30}$ and $\frac{9}{10}$ (c) $\frac{63}{72}$ and $\frac{49}{56}$ (d) $\frac{90}{120}$ and $\frac{45}{60}$

5. (a) $\frac{15}{25}$ (b) $\frac{9}{15}$ (c) $\frac{8}{16}$

F - 3 ADDITION

1. $40\frac{1}{4}$ 2. $26\frac{1}{6}$ 3. $36\frac{1}{16}$ 4. 50   5. $2\frac{3}{80}$ 6. $14\frac{43}{45}$

7. $27\frac{5}{12}$ 8. $18\frac{2}{3}$ 9. $12\frac{9}{10}$ 10. $146\frac{11}{12}$ 11. $33\frac{1}{8}$ 12. $20\frac{2}{15}$

13. $25\frac{5}{6}$ 14. $\frac{4}{5}$ 15. $31\frac{5}{18}$ 16. $142\frac{2}{3}$ 17. $2\frac{13}{15}$ 18. $20\frac{23}{40}$

19. 1   20. $17\frac{3}{4}$ 21. $22\frac{7}{8}$ 22. $1\frac{3}{4}$ 23. $1\frac{13}{15}$ 24. $27\frac{31}{40}$

25. $31\frac{23}{30}$ 26. $349\frac{3}{4}$ 27. $1\frac{1}{5}$ 28. $9\frac{7}{12}$ 29. $13\frac{19}{24}$ 30. $68\frac{4}{15}$

F - 4 SUBTRACTION

Set One:

1. $6\frac{3}{8}$ 2. $7\frac{5}{8}$ 3. $51\frac{3}{10}$ 4. $3\frac{5}{24}$ 5. $18\frac{7}{9}$ 6. $1\frac{2}{5}$

7. 0   8. $7\frac{2}{5}$ 9. $17\frac{9}{16}$ 10. $13\frac{1}{24}$

Set Two:

1. $4\frac{3}{4}$ 2. $3\frac{1}{4}$ 3. 0   4. $23\frac{1}{3}$ 5. $24\frac{2}{3}$ 6. $3\frac{1}{8}$

7. 1   8. $6\frac{4}{5}$ 9. 0   10. $\frac{7}{16}$

Set Three:

1. $3\frac{1}{4}$ 2. $9\frac{3}{4}$ 3. $12\frac{3}{8}$ 4. $13\frac{5}{8}$ 5. $\frac{7}{100}$ 6. $5\frac{5}{8}$

7. $1\frac{5}{6}$ 8. $28\frac{3}{4}$ 9. $6\frac{1}{5}$ 10. $\frac{1}{12}$

F - 5 MULTIPLICATION

Part One:

1. $2\frac{2}{3}$ 2. $\frac{21}{40}$ 3. $\frac{8}{9}$ 4. 72   5. 48   6. 2

7. 2   8. $3\frac{3}{4}$ 9. 6   10. 819

Part Two:

1. 2   2. 49   3. $8\frac{1}{2}$ 4. 2430   5. 6513   6. $\frac{1}{16}$

7. $8\frac{1}{16}$ 8. $4\frac{1}{5}$ 9. 90   10. $6\frac{1}{4}$

Part Three:

1. 6000   2. 66   3. $\frac{1}{8}$ 4. $8\frac{7}{25}$ 5. 44   6. 3

7. $5\frac{1}{3}$ 8. 252   9. 90   10. 1533

Part Four:

1. $\frac{17}{18}$ 2. $7\frac{5}{8}$ 3. 4   4. $1200.00   5. 2340 mi.   6. $\frac{1}{8}$

7. 13 lb.

F – 6 DIVISION

Set One:

1. 4	2. $10\frac{1}{2}$	3. $\frac{9}{10}$	4. 3	5. 5	6. 3
7. 36	8. 2	9. $\frac{1}{8}$			

Set Two:

1. 90	2. 325	3. $\frac{7}{12}$	4. 8	5. $\frac{5}{6}$	6. $7\frac{1}{3}$
7. $\frac{1}{36}$	8. 25	9. 24			

Set Three:

1. $\frac{1}{24}$	2. $1\frac{7}{25}$	3. $2\frac{13}{16}$	4. $1\frac{1}{3}$	5. 1	6. 30
7. $4\frac{1}{8}$	8. $\frac{3}{4}$	9. 5			

PROBLEMS WITH FRACTIONS

1. \$905 625 2. $9\frac{1}{2}$ in. 3. 20 4. 64 5. $45\frac{1}{2}$ cups
6. $7\frac{3}{8}$ lb. 7. $4\frac{1}{2}$ cups 8. (a) \$316.20 (b) \$270.00 (c) \$381.00
9. $1\frac{3}{4}$ ft. 10. $15\frac{5}{8}$ cups 11. 224

ANSWERS FOR DECIMAL EXERCISES

D – 3 CHANGING DECIMALS TO FRACTIONS

1. (a) $\frac{1}{5}$ (b) $1\frac{3}{5}$ (c) $2\frac{2}{25}$ (d) $16\frac{1}{400}$ (e) $\frac{1}{8}$ (f) $\frac{17}{100}$
   (g) $5\frac{7}{8}$ (h) $5\frac{7}{80}$ (i) $\frac{111}{125}$ (j) $95\frac{19}{20}$ (k) $\frac{91}{200\,000}$ (l) $200\frac{3}{40}$
2. (a) all (b) 0.214 and 0.2140 (c) $\frac{3}{4}$ and $\frac{15}{20}$ (d) $\frac{16}{100}$ and $\frac{160}{1000}$ (e) $\frac{150}{100}$ and $1\frac{1}{2}$
   (f) 2.22 and 2.220

D – 4 ADDITION AND SUBTRACTION

1. 684.348 2. 22.72 3. 1 4. 9.2 5. 37.464 6. 0.36
7. (a) 0.46 (b) 0.38 (c) 0.183 (d) 0.125
8. 7.41 9. 0.013 10. 100.8436 11. 18.75 12. 9.001
13. (a) *Vertical:* 29.46, 30.44, 73.12, 9.091; *Horizontal:* 26.75, 30.546, 84.815.
    (b) *Vertical:* 40.46, 129.5, 65.97, 118.175; *Horizontal:* 60.525, 52.65, 240.93.
       Total: (a) 142.111 (b) 354.105

D – 5 MULTIPLICATION

Set One:

1. 0.024	2. 288	3. 560	4. 0.15	5. 1920	6. 0.08
7. 120	8. 7652	9. 0.008	10. 6.25	11. 0.006	12. 4

Set Two:

1. 0.099	2. 0.3465	3. 652	4. 2700	5. 9
6. 0.000 012	7. 0.0081	8. 145.44	9. 0.54	10. 0.000 252
11. 1	12. 12.865			

D - 6 DIVISION

1. (a) 400   (b) 2250   (c) 1600   (d) 250 000   (e) 72   (f) 900
   Set A: 5   5   0.5   50 000   0.0005; *Sum:* 50 010.5005
   Set B: 0.5   0.5   0.005   0.05   5;   Sum: *6.055*
   Set C: 4   400   0.0004   400   4;   *Sum:* 808.0004
3. (a) 0.76   (b) 7.65   (c) 1.08   (d) 62.22   (e) 0.07   (f) 0.00   (g) 5.66   (h) 16.60
4. (a) 25   (b) 10   (c) 200   (d) 2000   (e) 0.5   (f) 100

D - 7 SHORT-CUT MULTIPLICATION AND DIVISION

1. (a) 360 000   (b) 0.000 19   (c) 0.625   (d) 3 475 000   (e) 0.62   (f) 25.6
   (g) 43   (h) 4.623   (i) 1
2. (a) multiply by 10^4 (b) multiply by 10   (c) multiply by 10^3
   (d) multiply by 10^2 (e) divide by 10^4

D - 8 ROUNDING

1. (a) 1.26   (b) 0.33   (c) 146.25   (d) 0.45
2. (a) 14   (b) 652   (c) 2   (d) 32 157
3. (a) 2.015   (b) 0.028   (c) 4.200   (d) 1600.251
4. (a) 0.083   (b) 0.219   (c) 0.778   (d) 0.733

D - 9 DECIMAL EQUIVALENTS

1. (a) 0.75   (b) 0.1   (c) 0.6   (d) 0.36   (e) 5.6875
   (f) 3.54
2. (a) 0.67   (b) 0.71   (c) 0.87   (d) 6.44   (e) 11.28
3. (a) 0.333   (b) 0.257   (c) 0.469   (d) 12.833   (e) 14.917

PRACTICE WITH DECIMALS (I)

1. (a) forty and seventeen ten-thousandths
   (b) two hundred sixty-five hundred-thousandths
2. (a) 17.007   (b) 0.805   (c) 800.005
3. (a) $3\frac{6}{25}$ (b) $1\frac{2}{125}$ (c) $\frac{1}{16}$ (d) $5\frac{19}{40}$ (e) $\frac{13}{3125}$ (f) $1\frac{14}{25}$
4. (a) 273.7825   (b) 61.284   (c) 74.4   (d) 200   (e) 0.002
   (f) 1   (g) 2.8   (h) 4   (i) 0.027   (j) 603.2
5. (a) 2.9   (b) $1.41\dot{6}$ (c) 6.2   (d) 9.175   (e) 0.714   (f) 0.594
6. (a) 0.16   (b) 1.6   (c) 5.064   (d) 170
7. (a) 0.0072   (b) 7.2   (c) 0.72   (d) 0.072
8. (a) 12   (b) 0.012   (c) 12 000   (d) 1200   (e) 0.000 12
   Sum: 13 212.012
9. 0.425   10. (a) 17.48   (b) 0.3726   (c) 4.5   (d) 200
11. 0.7°C   12. 0.01

PRACTICE WITH DECIMALS (II)

1. (a) two and one hundred fifty-six ten-thousandths
   (b) three thousand and forty-five thousandths
2. (a) 7.007   (b) 9000.0006   (c) 9006.010
3. (a) $4\frac{1}{40}$ (b) $62\frac{1}{12}$ (c) $\frac{1}{32}$ (d) $500\frac{2}{25}$
4. (a) 1   (b) 18.01   (c) 3.28
5. (a) 0.027   (b) 0.223 872   (c) 341 750   (d) 0.066   (e) 3.072
6. (a) 0.06   (b) 6000   (c) 0.06   (d) 0.06   (e) 0.06
7. (a) 14   (b) 6.4   (c) 0.02   (d) 1000.414
8. (a) 2   (b) 2.575   (c) 0.12   (d) 4   (e) 3.069

PROBLEMS INVOLVING DECIMALS

1. 16.6 L   2. 526.4 km   3. 0.3 cm   4. $2.66   5. $150.00   6. 100
7. 9 tenths, three hundredths, and 5 thousandths
8. 650, 16, 6.6, 6.56, 6.5, 5.6, 0.065   9. 0.675   10. $304

PRACTICE WITH PERCENT (I)

1. (a) 40%   (b) $\frac{1}{12}$ (c) 0.025   (d) $\frac{1}{8}$ (e) $16\frac{2}{3}\%$ (f) 0.875, $87\frac{1}{2}\%$
   (g) $\frac{7}{40}$, 17.5%   (h) $\frac{7}{200}$, 0.035   (i) $\frac{1}{80}$, $1\frac{1}{4}\%$ (j) 0.005, 0.5%
2. *Set A:* $45 + 30 + 24 + 0.8 + 48 = 147.8$
   Set B: $300 + 7.5 + 0.8 + 13 + 285 = 606.3$
   Set C: $8 + 629 + 2470 + 234 + 194.4 = 3535.4$
3. A: 4.8%   B: 2%   C: 4.24%   D: 6.12%   E: $8\frac{1}{3}\%$ F: 10%
4. (a) $133.33   (b) $10.00   (c) $5.33   (d) $330.00   (e) $3.00
5. (c) 60   (d) 70   (e) 25   (f) 20%   (g) 1.44

PRACTICE WITH PERCENT (II)

1. (a) 20%   (b) $\frac{1}{6}$ (c) 0.4375   (d) $\frac{7}{8}$ (e) $41\frac{2}{3}\%$ (f) 0.16, 16%   (g) $\frac{19}{20}$, 95%
   (h) $\frac{31}{400}$, 0.0775   (i) $\frac{1}{400}$, $\frac{1}{4}\%$ (j) 0.002, 0.2% (or $\frac{1}{5}\%$)
2. *Set A:* $480 + 30 + 1.6 + 150 + 1000 = 1661.6$
   Set B: $10.8 + 60 + 5 + 450 + 36.72 = 562.52$
   Set C: $1260 + 1 + 25 + 595 + 51.2 = 1932.2$
3. (a) $200.00   (b) $300.00   (c) $23.75   (d) $60.00   (e) $6.00   (f) $0.90
4. (a) 2400   (b) $71\frac{3}{7}\%$ (c) 50   (d) 20%   (e) 1.9   (f) 10   (g) 80%
5. (a) $16.80 + $26.04 + $15.12 + $29.40 = $87.36

PRACTICE EXERCISE: AREA OF PLANE FIGURES

1. (a) $A = s^2$; 0.16 m²   (b) $A = lw$; 3.84 cm²   (c) $A = \pi r^2$; 3.14 sq. ft.

    (d) $A = \dfrac{bh}{2}$; 57 sq. in.   (e) $A = \pi r^2$; 154 sq. in.   (f) $A = lw$; 0.02 m²

    (g) $A = h\dfrac{b_1 + b_2}{2}$; 40 sq. in.   (h) $A = \dfrac{bh}{2}$; 270 sq. in. or $1\frac{7}{8}$ sq. ft.

    (i) $A = \pi r^2$; 113.04 cm²   (j) $A = s^2$; 1.21 m²   (k) $A = bh$; 42 sq. in.

    (l) $A = h\dfrac{b_1 + b_2}{2}$; 10.5 cm²

2. A. (a) square   (b) $A = s^2$ (c) $12\frac{1}{4}$ sq. ft.

    B. (a) parallelogram   (b) $A = bh$ (c) 112 sq. in.

    C. (a) circle   (b) $A = \pi r^2$ (c) 78.5 m²

    D. (a) trapezoid   (b) $A = h\dfrac{b_1 + b_2}{2}$ (c) 375 sq. in. or $2\frac{29}{48}$ sq. ft.

    E. (a) triangle   (b) $A = \dfrac{bh}{2}$ (c) 0.21 m²

    F. (a) rectangle   (b) $A = lw$ (c) 675 sq. in. or $4\frac{11}{16}$ sq. ft.

PRACTICE EXERCISE: SOLVING EQUATIONS

Set One: Values of y: 1. 5   2. 48   3. $11\frac{1}{2}$ 4. 6.3   5. $1\frac{1}{16}$

Set Two: Values of k: 1. 8   2. 7   3. 2   4. 0.1   5. 5

Set Three: Values of a: 1. 96   2. 3.2   3. 15   4. 3   5. 35

Set Four: Values of m: 1. 7   2. 9   3. 20   4. 2   5. 3

Set Five: Values of n: 1. 5   2. 6   3. 12   4. 1   5. 24

Set Six:

1. $c = a + b - k$ 2. $k = \dfrac{ac}{b}$ 3. $b = \dfrac{ck}{a}$ 4. $c = ab - k$

5. $a = \dfrac{k}{bc}$ 6. $b = ck - a$ 7. $a = \dfrac{k - c}{b + c}$ 8. $b = k + \dfrac{c}{a}$

9. $a = k - bc$ 10. $c = \dfrac{k - a}{b}$